ECONOMIC COMMISSION FOR EUROPE
Geneva

SOFTWARE FOR
INDUSTRIAL AUTOMATION

UNITED NATIONS
New York, 1987

ECE/ENG.AUT/29

UNITED NATIONS PUBLICATION
Sales No. E.87.II.E.19
ISBN 92-1-116392-7

03500P

STUDY ON SOFTWARE FOR INDUSTRIAL AUTOMATION

CONTENTS

GE.87-21217 /2892d, 2896d

List of tables

List of figures

The present study has its origin in a decision taken in March 1984 by the Working Party on Engineering Industries and Automation, a subsidiary body of the Economic Commission for Europe, to include in its programme of work a study on recent developments in software means for industrial automation.

The first ad hoc Meeting for the study, held in June 1985, defined the scope of the study to be undertaken in relation to other automation-oriented projects carried out by the Working Party. The Meeting agreed "that the study should focus on software for discrete manufacturing processes, in particular software for automation in engineering industries". It identified the problems of common interest to be considered in the study and agreed on the outline. It was decided that the secretariat should send a circular letter to all ECE member countries and relevant international organizations inviting them to nominate rapporteurs and requesting submission of material relevant to the study. In response to that letter, the Governments of Finland, Ireland and the Netherlands nominated national rapporteurs and Czechoslovakia, the United Kingdom, the USSR and the United States submitted valuable material for the preparation of the draft study. The Council for Mutual Economic Assistance (CMEA), the International Organization for Standardization (ISO), the International Electrotechnical Commission (IEC) and the World Intellectual Property Organization (WIPO) also contributed to the draft study.

At its sixth session, in March 1986, the Working Party endorsed the Report of the first ad hoc Meeting for the study and requested the secretariat to proceed with the preparation of the study and to circulate it to national rapporteurs for comments, to convene a second ad hoc Meeting and to modify the title of the study to read "Software for Industrial Automation".

The first draft of the study was discussed at the second ad hoc Meeting, convened in November 1986. After an in-depth consideration of the draft, the Meeting suggested various amendments and revisions, and agreed on the schedule for the finalization of the project.

The Working Party, at its seventh session, in February 1987, took note of the work undertaken, endorsed the proposed schedule for the finalization of the study and decided to issue the English and French versions of the study as sales publications and the Russian version as a regular ECE derestricted document.

The final version of the study has been prepared by the secretariat on the basis of the comments and instructions of the second ad hoc Meeting for the study, the comments received from national rapporteurs, by incorporating additional material and comments received from Bulgaria, France, the Netherlands and the United Kingdom and making use of the advice kindly provided by the Chairman and Vice-Chairman of the two ad hoc Meetings for the study, Mr. R. Ahlberg (United States) and Prof. D. Kochan (German Democratic Republic), respectively.

The present study is one in a series dealing with industrial automation which have either been undertaken or which are in the process of preparation within the framework of the programme of work of the Working Party on Engineering Industries and Automation. At the beginning of 1985, a study entitled Production and Use of Industrial Robots was published (United Nations publication, Sales No. E.84.II.E.33). As a follow-up to that study, the ECE Seminar on Industrial Robotics '86 - International Experience, Developments and Applications was held at Brno (Czechoslovakia) from 24 to 28 February 1986. For further information concerning the Seminar, reference is made to the respective Report (ENG.AUT/SEM.5/4). The beginning of 1986 saw the publication of a study entitled Recent Trends in Flexible Manufacturing (United Nations publication, Sales No. E.85.II.E.35).

This study incorporated the main findings of another ECE Seminar entitled "Flexible Manufacturing Systems: Design and Applications" which was held in Sofia (Bulgaria) from 24 to 28 September 1984 at the invitation of the Government of Bulgaria. The report of this Seminar bears the symbol ENG.AUT/SEM.3/4.

Other studies dealing with industrial automation currently being undertaken by the Working Party include:

- Trends in the electrical and electronics industries; and

- Technical achievements in telecommunication equipment and implications for industry.

Occasionally, the Working Party also deals with automation in non-industrial sectors. Recently, a major study on innovation in biomedical engineering was released, illustrating current developments and trends in Digital Imaging in Health Care (Sales No. E.86.II.E.29).

In the preparation and finalization of the present study, the Economic Commission for Europe wishes to express its thanks to Governments, international organizations, and individual specialists who have provided material and given the secretariat the benefit of their experience and advice without which it would not have been possible to finalize the study.

The ECE secretariat's thanks go, in particular, to Prof. D. Kochan, Technical University of Dresden (German Democratic Republic) who assisted the secretariat in the preparation of the first draft of the study.

Note: Mention of any manufacturing company or other institution in the context of the present study does not imply endorsement by the United Nations.

LIST OF ABBREVIATIONS FREQUENTLY EMPLOYED IN THE STUDY

AC	Adaptive control
ACC	Adaptive control constraint
ACO	Adaptive control optimization
ACS	Adaptive control system
ADAPT	Adaption of APT
AE	Acoustic emission
AGV	Automated guided vehicle
AI	Artificial intelligence
ANSI	American National Standards Institute (United States)
APT	Automatic programming for tools
AS/RS	Automated storage and retrieval system
ATS	Automated time standards
BTR	Behind the tape reader
CAD	Computer-aided design
CAE	Computer-aided engineering
CAM	Computer-aided manufacturing
CAM-1	Computer Aided Manufacturing-International (International Organisation)
CAP	Computer-aided planning
CAPP	Computer-aided process and production planning
CL	Cutter location
CIM	Computer-integrated manufacturing
CIR	Computer-internal representation
CLDATA	Cutter location data
CMEA	Council for Mutual Economic Assistance
CNC	Computerized numerical control
CP	Continuous path (control)
CPU	Central processing unit
DNC	Direct numerical control
ECE	Economic Commission for Europe
EDP	Electronic data processing
EEC	European Economic Community
EMO	European Machine Tool Exhibition
ES	Expert system
FMC	Flexible manufacturing cell
FMS	Flexible manufacturing system
FMU	Flexible manufacturing unit
GKS	Graphical kernel system
IC	Integrated circuit
ICAM	Integrated computer-aided manufacturing
IEEE	Institute of Electrical and Electronic Engineers (United States)
IEC	International Electrotechnical Commission

IFAC	International Federation of Automatic Control
IFIP	International Federation for Information Processing
IGES	Initial graphic exchange specification
I/O	Input/output
IR	Industrial robot
ISO	International Organization for Standardization
IT	Information technology
ITV	Industrial television
LAN	Local area network
MAP	Manufacturing automation protocol
MC	Machining centre
MIS	Management information system
MPST	Modular multiple processor controlling system
MRP	Material requirements planning
MPS	Modular program system
MTC	Machine-tool controller
NC	Numerical control
OSI	Open systems interconnection
PC	Programmable controller
PDES	Product definition exchange standard
PDDI	Product data definition interface
PMC	Programmable machine controller
PTP	Point-to-point (control)
R and D	Research and Development
ROM	Read-only memory
SET	System for exchange and transfer
SME	Society for Manufacturing Engineers (United States)
STEP	Standard for exchange product definition data
TOP	Technical and office protocol
TTL	Transistor transistor logic
UNESCO	United Nations Educational, Scientific and Cultural Organization
WIPO	World Intellectual Property Organization

I. GENERAL INTRODUCTION, BASIC DEFINITIONS AND OBJECTIVES OF THE STUDY

I.1 Introduction

The process of industrialization, which started with the invention of the steam engine, brought about a tremendous development in technology and, consequently, productivity. The improvement of technology generally consisted of a refinement of mechanical aids in the production process. The mechanization of manual work with the help of several kinds of machines and simple mechanical and electro-mechanical control devices, relays, mechanical controllers, etc. culminated in the introduction of the assembly line. As the traditional assembly line combined the advantage of very high productivity with the disadvantage of extreme rigidity, the application of this technology was limited to mass production processes.

With the demand for mass-produced manufacturing goods starting to decline and customized production becoming commonplace in many manufacturing industries, engineering industries have had to comply with the requirements of medium-to-small-batch series of products and the growing demand for flexibility. In order to meet this need, the conflict between flexibility and productivity had to be overcome. The key to the solution of this problem was the digitalization of control processes in manufacturing and the integration of computers in technological processes. Changes in technological sequences of operation, which previously had involved major changes in technical hardware (resulting in high costs and long set-up times), became easier to make through changes in the control software manipulating the technical hardware (machine-tools with NC, CNC, DNC; robots; FMS; etc.).

Progress in computerized manufacturing was originally dependent on advances in information technology hardware (cost, speed, information storage capacity, reliability of data processing, etc.). It was with the advent of the microprocessor that this situation changed considerably. As most of the constituent control hardware for computer-aided manufacturing systems became available and tended to become more powerful, more reliable and less costly, most interest became focused on developments in software and communication aspects.

In the early 1950s the automation process in manufacturing started at the level of the single machine tool with the application of numerical control. The development did not stop with the integration of one or more machine tools with their related auxiliary support devices, such as material-handling systems, robots, tool-handling systems and machines performing secondary operations into machining cells. The logical follow-up, which can be noted currently in many manufacturing shops, is the integration of a number of machining cells into flexible manufacturing systems [8].

The challenge of the future will be to utilize computer technology in the manufacturing process from the moment of product conception according to market information to its final delivery to the customer [10]. The success of this concept of computer-integrated manufacturing (CIM) will depend, to a large extent, on the capability and reliability of the information system which controls and supervises all processes. The traditionally relatively independent functioning of information processing and material processing will have to be merged into a single interdependent complex.

The ability to integrate and link all elements together will depend on the development of software to make the various components communicate with one another and to control their "co-operative functioning" [3].

The complexity of tasks to be performed by appropriate software for a CIM system as well as the role of the central data base and an effective systems architecture are described in Figure 1. CIM is a concept that relies on a common manufacturing data base and a clearly structured information system that effectively links all main functions of the factory of the future: engineering design, manufacturing planning and control and factory automation.

The key problem of any integration is the ability to communicate. Specific communication needs have to be met through adequate communication architecture. In this context, it is generally recognized that local area networking (LAN) technology will eventually form the backbone of communication in CIM [8]. However, the selection of an effective systems architecture and effective information channels do not solve the communication problem completely. Standardized communication protocols are of decisive importance, if communication between "islands of automation" is ever to work properly. The development of standards for interfaces and modular control systems for the exchange of information must constitute priority areas for research and development in order to set up compatible systems containing hard- and software from a variety of manufacturers [6]. The activities of the International Organization for Standardization (ISO) are of particular significance in this process. As a result of international co-operation between national standards bodies, an International Standard Reference Model of Open Systems Interconnections (OSI) was approved in 1984 [11]. This is intended to provide a common basis for the co-ordination of standards development for the purpose of systems interconnection.

While originally interest was focused on the process of industrial automation at the level of the shop floor, in recent years this interest has shifted towards the domain of engineering, production planning and administrative control. As a result of the fundamental changes which have taken place in the character of work, not only at the shop floor, but also in engineering, production planning and administration, many economic, social, qualification, training and retraining problems have arisen which have to be solved.

I.2 Scope and objectives of the study

The aim of the present study is to highlight the increasing importance of software for industrial automation. The study surveys recent developments in automation of discrete manufacturing processes, in particular in engineering industries, and the key role played by software in several technological applications. It should provide assistance in the consideration of trends in the development and application of software for industrial automation with a

Figure 1.   Model of a CIM system [21]

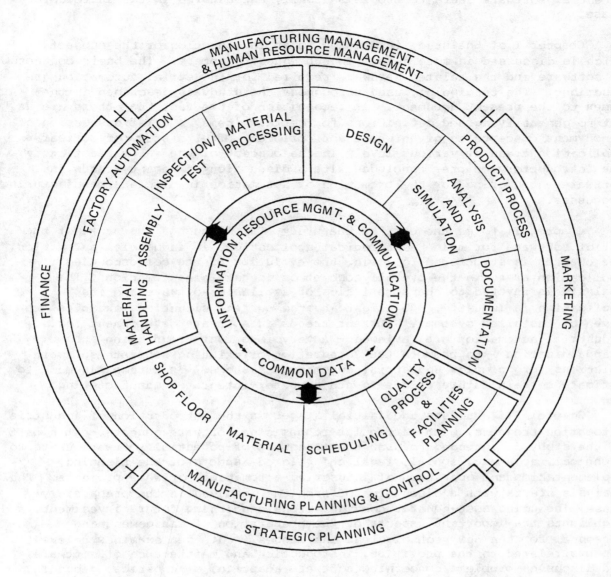

view to estimating business opportunities and drawing the attention of potential software users to problems usually encountered in the introductory phase.

Chapter I of the study is devoted to an introduction of the subject, which is discussed in more detail in subsequent chapters. The basic concepts of software and the related hardware required for industrial automation are described. The terminology used throughout the study is described in the annex to the present document. In recognition of the fact that there are as yet no generally agreed definitions for certain items, a brief discussion of synonymous concepts is given where appropriate. After a study of software application areas at various levels in the concept of the automated factory, the introductory chapter concludes with a discussion of recent trends in software engineering for the promotion of automation in discrete manufacturing processes.

Chapter II forms the core of the study. It discusses the state of the art of software for main technological applications in industrial automation. The chapter is structured following the evolution of the corresponding application areas in the overall concept of industrial automation. The analysis is devoted to characteristics of available software and its application in industry. To illustrate the current situation, examples are given of installed systems. Current trends in software development for industrial automation are reviewed - a very important trend being standardization - to promote the integration of existing separate software solutions into complex industrial automation systems. An attempt is made to estimate certain future trends toward computer-integrated manufacturing.

Chapter III analyses the initial stages in the introduction of industrial automation projects, with particular emphasis on software aspects. This part of the study is intended to draw the attention of potential software users to technical, economic, social, legal and related aspects of the planning, implementation and operation of industrial-automation-software projects. The analysis starts with a discussion of problems involved in the preparatory phase, including cost-benefit considerations concerning future investment, and highlights the important aspects of the preparation of management and personnel for the new technological challenges. It then examines relevant aspects related to the provision, introduction and maintenance of software, and discusses problems concerning software adaption, diagnostics, debugging and maintenance in the implementation and operational phases.

Chapter IV reviews national and international programmes intended to facilitate the development and introduction of advanced software for industrial automation. It includes an examination of relevant governmental programmes launched in several ECE countries and in Japan and of activities of relevant international governmental and non-governmental organizations actively promoting international co-operation in the field under consideration. Although the chapter focuses on software-related aspects of measures to promote industrial automation, it has been considered useful to include in the survey selected activities of governments and international organizations in the broader range of promotion of industrial automation and the application of information technology in engineering industries in general - which in any case includes software solutions.

Finally, an attempt is made to draw some general conclusions regarding the planning, implementation, operation and maintenance of industrial automation software.

I.3 Definitions

Data processing and industrial automation terminology, which in some cases is not yet based on internationally agreed definitions, is used extensively throughout the present study. Attempts have been made to compile definitions of those terms from several international sources. Those definitions are included in annex I. For many items, several definitions currently in use have been found worthy of inclusion in this list. Although alternative definitions vary in their wording, they coincide in most cases as regards the philosophy behind the concept. The definitions are intended to serve as a basis for understanding certain key concepts referred to in the study. They will be cross-referenced at appropriate places in the text.

I.4 Software at various levels in the concept of the automated factory

Independent of the degree of the mutual interlinking and, therefore, independent of the level of automation achieved, computer support is possible in all levels of an operational hierarchy and, thus, specific applications software is necessary.

In accordance with the productivity pyramid described by Allen-Bradley [25] and by generalizing further development, the following main levels of operational hierarchy may be identified:

1.   Plant level

   Responsible for overall production scheduling and execution. Requires two-way communication between mainframe computer and lower levels.

2.   Medium level

2.1.   Organizational tasks

   Schedules production and provides management with information by monitoring and supervising lower levels.

2.2.   Technical tasks

   All engineering-related tasks for the structural design of products, units and single parts (CAD functions, such as drafting, designing, computing, documentation, and including automatic drawing) and the tasks comprised in technological production

- 9 -

planning (CAP functions, such as elaboration of operations and
their sequences, NC-programming, updating service, determination
of time and costs).

3. Cell level

Co-ordinates the production flow amongst various stations.
Integrates individual stations into an automated system.

4. Station level

Performs the logic necessary to convert the input from lower
levels into output commands, based on hierarchic directives.
Process computers for realizing DNC modes (see chapter II.1.1) are
also usually integrated into this level.

5. Machinery/Process level

Basic interface (on-line or off-line) with plant floor equipment.
Includes NC-machine tools, manufacturing cells, FMS, industrial
robotics, equipment for transport and handling, storage and
others, and also devices and equipment for process monitoring,
sensorics and in-process measurement.

I.5 Recent trends in software engineering

Industrial automation systems represent complex software solutions. For
the development of such systems, it is necessary to ensure comprehensive and
detailed scheduling of the total project, including the application of
knowledge and rules summarized in software engineering.

As a branch of engineering research, software engineering establishes the
methodological framework, and supplies the methods and tools for the efficient
development, application and maintenance of software. Software engineering
covers the whole lifetime of a program, starting at the time the order is
received and continuing until the program is replaced by a new one. This
period embraces the so-called life cycle and consists of various steps
(figure 2):

1. Systems definition

2. Functional design, specification (determination of the problem,
functional design of the system, program-oriented design)

Functional design uses a computer-independent formal language
(design language) for the determination and elaboration of the
sequences. The result is an input for coding, independent of the
programming language.

3. Development

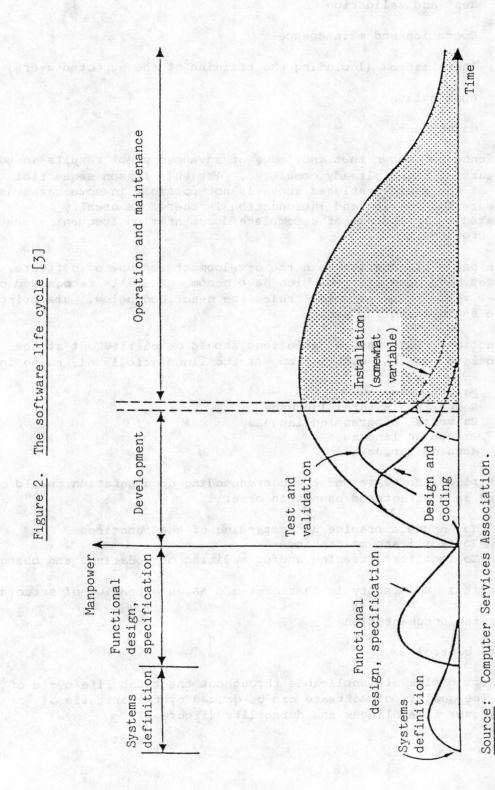

Figure 2. The software life cycle [3]

Source: Computer Services Association.

3.1   Design and coding

3.2   Test and validation

4.   Operation and maintenance

4.1   Installation (including the training of the expected users)

4.2   Application

4.3   Maintenance

Experience has shown that knowledge of advanced steps results in editing the statements of steps already completed. For this reason sequential processing of the steps mentioned above is not possible in some cases and the time necessary for testing and introduction is therefore usually underestimated. The absence of a complete documentation frequently creates additional problems.

On the basis of experience in the development and use of software, various rules have been created which have become generally recognized over the past few years. The essential rules are described below. The principles of software engineering include:

-   Abstraction: All steps of solutions should be carried out at the appropriate level of abstraction At the linguistic level, these include:

    (a)   Natural language
    (b)   Restricted natural language
    (c)   Universal programming language
    (d)   Assembler language
    (e)   Machine language

-   Structuring: Software and the corresponding documentation should be drafted in a structured manner in order:

    (a)   To ensure a precise understanding of the functions
    (b)   To facilitate maintenance
    (c)   To facilitate tracing and/or auditing of updatings and changes

-   Modularity: Modularity is characterized as an extension of structuring

-   Integrated documentation

-   Quality control

These principles are applicable throughout the total life cycle of software. The quality of software can be gauged by the criteria of efficiency, user friendliness and durability (figure 3).

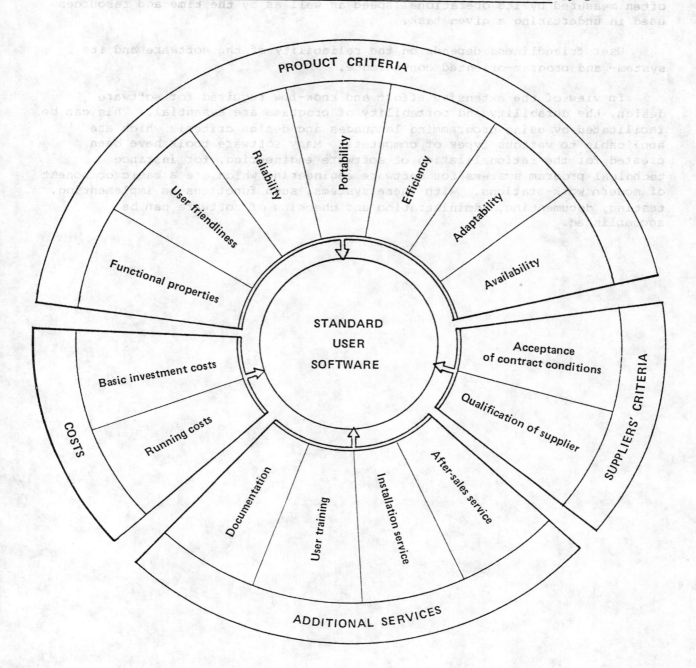

Figure 3. Essential criteria for standard user software [58]

Software efficiency depends on the skill of the development staff. It is often measured by its operational speed as well as by the time and resources used in undertaking a given task.

User friendliness depends on the reliability of the software and its system- and program-oriented convenience.

In view of the extensive effort and know-how required for software design, the durability and portability of programs are essential. This can be facilitated by using programming languages and design criteria which are applicable to various types of computers. Many software tools have been created for the rationalization of software engineering, for instance technical program systems for software engineering which are a basic component of modern work-stations. With these systems, such functions as implementing, testing, documenting, administration and checking of software can be accomplished.

II. PRESENT STATE AND CURRENT TRENDS IN SOFTWARE FOR INDUSTRIAL AUTOMATION

II.1. A survey of the availability and application of software for industrial automation

II.1.1. NC-machine tools, CNC, DNC, ACS

Numerically controlled (NC) machine tools

The machine tool with its control device constitutes the main part of any automated production plant. Since the late 1950s, automatic sequencing of the relative motions between workpiece and tool (including positioning and path measuring) have been possible for the main cutting processes, i.e. turning, drilling and milling.

Numerical control (for definition, see annex I) performs the function of direct control of the technical equipment in the production-technological basic system, in particular in small and medium-batch production.

With respect to functions undertaken by the control device, two essential principles can be distinguished:

- Functional conversion of the input information by hard-wired program controls (NC); and

- Functional conversion of the input information by free-programmable controls (CNC).

The two essential stages in the development of the numerical control of machine tools may be outlined as follows:

- Hard-wired program controls (relay, transistor, TTL);

  . machine-related controls;

  . simple numerical control; and

- free-programmable controls (minicomputer, microprocessor).

The hard-wired program controls are characterized by a special sub-assembly which is assigned to each functional block for the following tasks:

- Decoding of the input information of the NC-program;

- Conversion of this information into path and switching commands and their output to the control elements of the machine tool;

- Interpolation with respect to the different numbers of controlled axes; and

- 15 -

- Feedbacks from the path measuring system of the machine and comparison with specified nominal values.

Thus, the major part of the work in the development of a hard-wired program control consists of the design of the special circuit and its realization by means of special components, the functional conversion being done mainly on the basis of hardware.

The necessary software, which is mostly deposited in punched tapes, is confined to the input information for the control system, i.e., the NC machining program prepared by the user. It contains both geometrical information for the tool path to be executed along the workpiece contour and the pertinent technological information, such as feed, cutting speed, cutting depth, and the tools to be used with their parameters.

The operating principle of NC is shown in figure 4.

Computerized numerical control (CNC)

Since the mid-1970s, there has been a clear shift in the focus of machine producers and users towards the application of computer controls for machine tools. At the beginning of this period, the core of these controls still consisted of minicomputers, but already at the fifth European Machine Tool Exhibition (EMO) in Paris, in 1983, almost all the exhibited controls of a higher level of development were based on the 16-bit microprocessor technique [26]. Developments currently under way (in particular in complicated cutting prcocesses) point to the application in the not-so-distant future of 32-bit microprocessor controls.

The functions which are realized with NC by hardware (circuit) are now taken over completely by the software complexes which are implemented in - mostly several - microcomputers (see also definition in annex I). The logic unit of CNC generates the control commands for the control elements of the machine from the input information of the NC machining program and the commands of the control algorithm which are mostly stored in read-only memories.

The main work of the development of control for CNC lies in the software development for the control algorithm. The programs implemented in the computers are subdivided into (see figure 5):

- The functional software, realizing the NC-specific functional content of the control (with pertinent internal main memories and data fields);

- The application software, representing the input information to be run for the functional software; and

- The service software, in general used for supporting the operations of starting, diagnosis and maintenance.

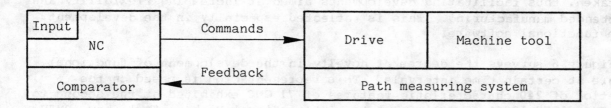

Figure 4. Principle of numerical control (NC) [23]

Input		Commands	Drive	Machine tool
NC				
Comparator		Feedback	Path measuring system	

Figure 5. Subdivision of the software complexes
implemented in CNC [23]

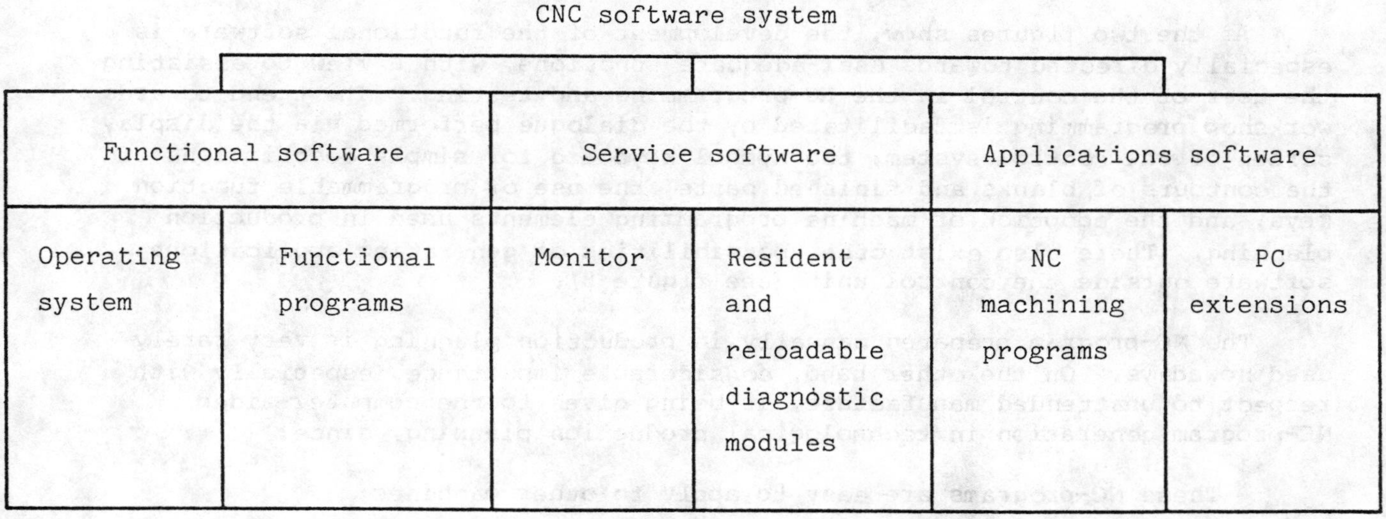

CNC software system

Functional software		Service software		Applications software	
Operating system	Functional programs	Monitor	Resident and reloadable diagnostic modules	NC machining programs	PC extensions

The widespread application of CNC, allows an extensive and diversified spectrum of tasks concerning the automation of the manufacturing process to be undertaken, thus facilitating developments aimed at increasing flexibility and low-attended manufacturing. This is reflected especially in the development of the functional software.

Figure 6 surveys the degree of novelty in the development of functional software at certain time intervals. This representation is based on the evaluation of 240 characteristic features of 71 CNC exhibited at the second to sixth EMO (1977-1985) as well as at the International Machine Tool Show (IMTS) in 1984.

In the development of free programmable control over the past ten years, three essential stages of development have taken shape as summarized in figure 7.

As the two figures show, the development of the functional software is especially directed towards user-adequate functions, with a view to assisting the user of the control in the NC-programming and testing. The trend towards workshop programming is facilitated by the dialogue performed via the display screen of the control system, the symbol keyboard for simple descriptions of the contours of blanks and finished parts, the use of programmable function keys, and the adoption of machine programming elements used in production planning. There also exist other possibilities of generating applications software outside the control unit (see figure 8).

The NC-program prepared manually in production planning is very rarely used nowadays. On the other hand, considerable importance, especially with respect to unattended manufacture, is being given to the computer-aided NC-program generation in technological production planning, since:

- These NC-programs are easy to apply to other machines;

- There is access to central data banks; and

- Proceeding from the NC-program, the information co-ordination must be carried out with a large number of programs for other process-accompanying automated functions (measuring, supervision, handling, transport).

In this context, programming systems for production planning have also been designed in a more user-adequate way. The functional software of CNC has been made consistent with machine programming systems, especially with respect to:

- Guiding of operators in conversational mode;

- Graphic support; and

- Symbolic programming and other routines to aid programming.

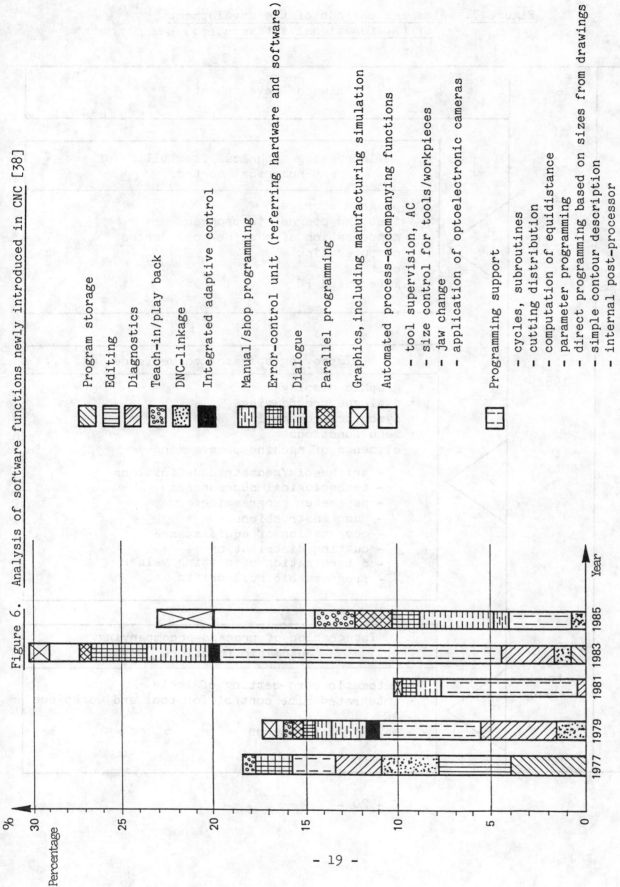

Figure 6. Analysis of software functions newly introduced in CNC [38]

Program storage
Editing
Diagnostics
Teach-in/play back
DNC-linkage
Integrated adaptive control

Manual/shop programming
Error-control unit (referring hardware and software)
Dialogue
Parallel programming
Graphics, including manufacturing simulation
Automated process-accompanying functions

- tool supervision, AC
- size control for tools/workpieces
- jaw change
- application of optoelectronic cameras

Programming support

- cycles, subroutines
- cutting distribution
- computation of equidistance
- parameter programming
- direct programming based on sizes from drawings
- simple contour description
- internal post-processor
- arithmetic/geometric specifications
- jump instructions
- menu
- selection of cutting values

(71 controls with 245 features)

(Based on: Evaluation of EMO 1977, 1979, 1981, 1983, partly 1985, IMTS 1980, 1984)

- 19 -

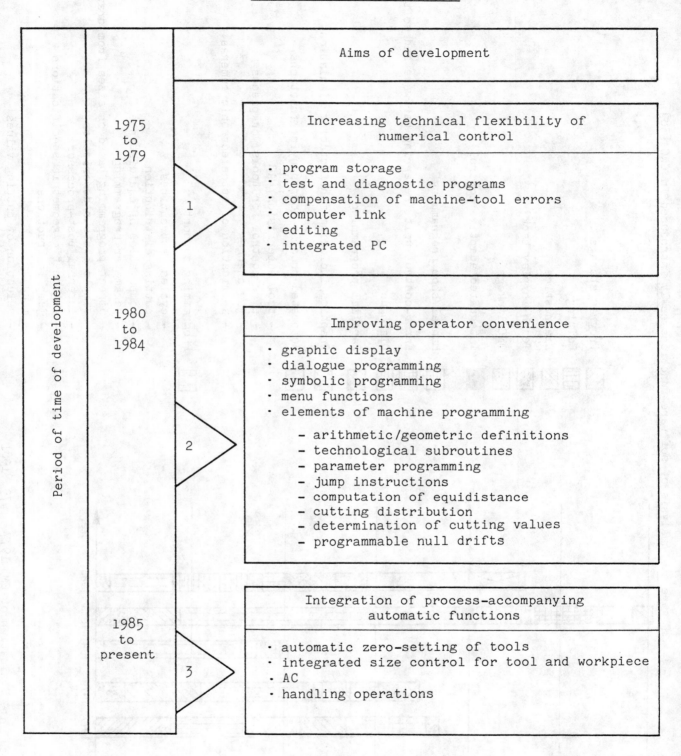

Figure 7. Aims and periods of the development of CNC functional software [38]

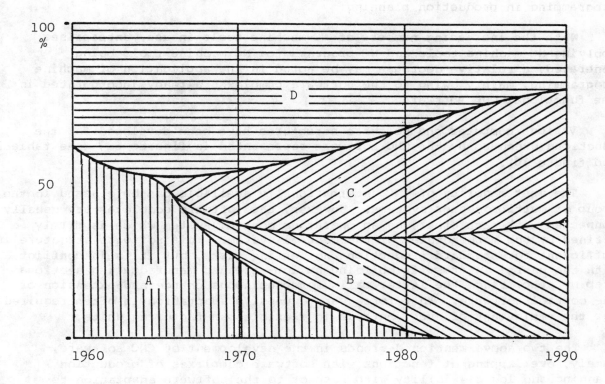

Figure 8. Applied methods of NC-programming in percentage [38]

A - Manual programming in production planning

B - Input of NC-program at the machine tool

C - Programming station with microcomputer

D - Programming station permitting access to central computer

This has resulted in considerable overlapping of functions (figure 9), so that essential software complexes remain unused especially by users of machine programming in production planning.

With the increasing numbers of NC-machine tools in use, enterprises applying NC-machine tools and NC-programs for NC-machine tools being generated, a relative decrease may be noted in the application of machine programming, mainly owing to the software complexes already incorporated in the functional CNC software.

Various computer and software structures have been developed for the functions executed by the CNC software represented in figures 6-7 (see table 1 and figure 10).

The vast majority of CNC internationally available can be classed in the group of multiprocessor systems. However, the software complexes are usually connected with each other, by an operating control system which is firmly defined by the manufacturer of the control system. This software structure is sufficient for standard production engineering tasks, but it is in conflict with the users' interest in building up low-attended manufacturing sections step by step, since the latter calls for continuous change and extension of the control software. Therefore, very expensive adaptation is often required for the employment of such controls in flexible manufacturing units.

The two above-mentioned trends in the development of CNC software, namely, overlapping of functions with software complexes of production planning and low flexibility with respect to the software adaptation to unmanned manufacturing owing to its firmly defined structure, give rise to a demand for modularly designed - and therefore functionally combinable - control systems.

In this respect, a number of feasible solutions have been worked out by the MPST 1/ working group in the Federal Republic of Germany. The aim of that working group is to create a control system which - owing to the increasingly economical use of microprocessors - allows also the numerical control of machine tools which can perform functions in addition to the standard manufacturing tasks (turning, drilling, milling) [27-32].

With respect to the software structure, a distinction is made between so-called active and passive modules, the passive ones (e.g. "output of speed values") being supplied by the active ones (e.g. "interpolation and position control").

The basic principle of modular software design consists in the formation of delimitable functions and the assignment of a self-contained functional block to each microprocessor for processing. The following functional blocks have been worked out [28]:

1/ MPST = Modular multiple processor controlling system.

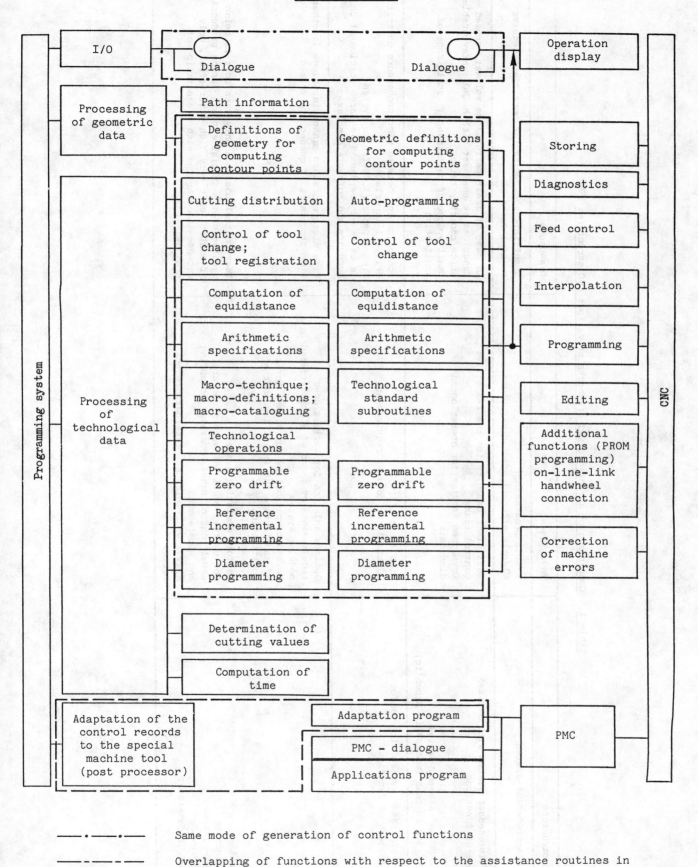

Figure 9. Overlapping of functions in machine programming systems and CNC [38]

——·——·—— Same mode of generation of control functions

——————— Overlapping of functions with respect to the assistance routines in the generation of programs

—— —— —— Partial overlapping of functions (PMC performs other tasks in addition)

Table 1. Comparison of advantages and drawbacks of various CNC structures [38]

CNC structure	Advantages	Drawbacks
Single-processor system	Simple programming Clearly arranged program structure	Insufficient capacity for more comprehensive control tasks
Multiprocessor system with a fixed range of functions Examples: - FANUC and Siemens series - CNC 600 (German Democratic Republic)	Sufficient capacity for the respective control tasks Developed service for operation and diagnosis	In case of great performance volume: unused functions at the user with MACRO, unfavourable price/capacity ratio Small volume of performance: small rage of application
Multiprocessor system with random assignment of functions	High flexibility Adaptability to user requirements	Communication between microprocessors constitutes critical interfaces High expense of operating system In cases of low demand, a large portion of unused hardware and software complexes
MPST Examples: - Procontic - IBH-control - Micromodus - EPM/Gildemeister	Optimum solution for various user requirements by means of a standardized modular system Upward compatibility for higher forms of manufacture (MC,FMS): low cost of modification Uniform operating system For users: equal software structure, equal control panel, operator prompting and programming Always well-balanced price/performance ratios	Free microprocessors cannot intervene in other microprocessors if queues are formed User control possible only when unit is being manufactured; initial expense during production of control unit very high (thereafter, only expense is connected with organization of production)

Figure 10. Overview of some CNC structures [38]

Kind of structure	Model structure (exemplary)	Character number of ppr	Number of functions	Explanation
Single processor systems	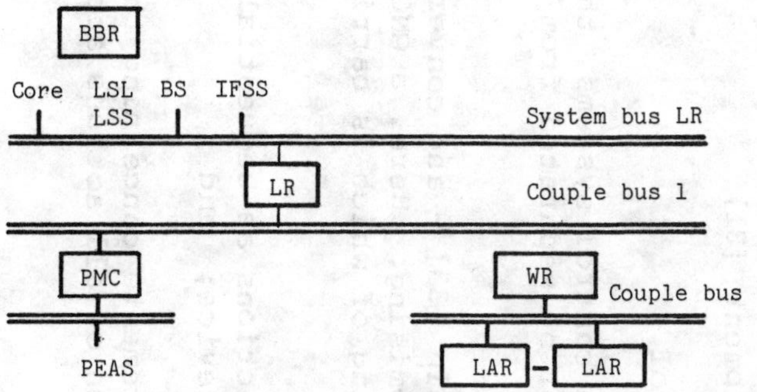	1	constant	A maximum of functions is to be guaranteed by a minimum of ppr.
Multiple processor systems	Determined functional volume	constant	constant	The present function is assigned to the unoccupied ppr.
Modular multiple processor controlling	Optional function assignment 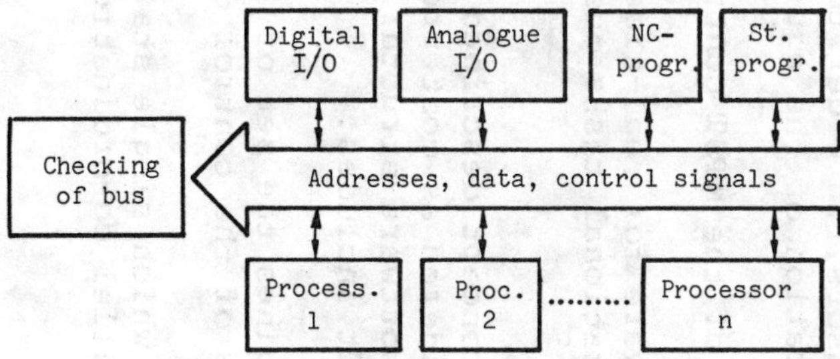	constant	varying	Functions which are unambiguously assigned to the ppr.
Modular multiple processor controlling system (MPST)	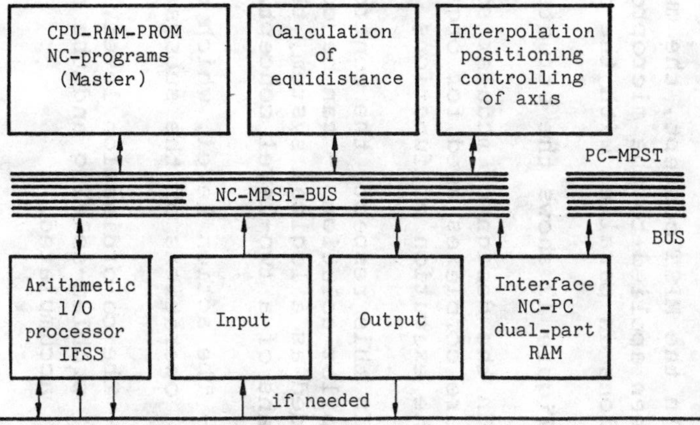	varying	varying	

Model structure (exemplary) diagrams include the following labels:

Multiple processor systems: Core, LSL, BS, IFSS, LSS, BBR, LR, PMC, PEAS, System bus LR, Couple bus, WR, Couple bus 1, LAR, Digital I/O, Analogue I/O, NC-progr., St. progr., Processor n, Proc. 2, Process. 1, Addresses, data, control signals, Checking of bus.

MPST diagrams include: CPU-RAM-PROM NC-programs (Master), Calculation of equidistance, Interpolation positioning controlling of axis, Arithmetic I/O processor IFSS, Input, Output, Interface NC-PC dual-part RAM, NC-MPST-BUS, PC-MPST BUS, if needed.

Legend:

BBR – Operating station
BS – Display
IFSS – Interface for star-connected instruments with serial information transmission
LAR – Position controller

LR – Master computer
LSL – Punched tape reader
LSS – Punched tape punch
ppr – Microprocessor

PEAS – Process input-output-control
PMC – Programmable machine controller
St-progr. – Control program
WR – Path controller

- Operating and controlling data - input/output;

- NC data management, processing and distribution;

- Geometric information processing;

- Technological (programmable controller) function; and

- Centralized control.

To each functional block are allocated so-called "functions which may be entrusted with orders", by which the content of a functional block is defined more specifically (figure 11).

For all software components a standard interface has been proposed. Since the software has to be adapted to the specific user environment, computer technology is available to the effective renewal of the functional software for an MPST control [32].

In the MPST concept, the multitasking principle of the process computers has been applied to the microprocessors, in order to ensure execution of functions in parallel with the operation of the equipment [31].

Figure 12 shows the structure of the MPST control.

In the design of modular software for real-time control systems, the software complexes used for organizational tasks must be separated from those for the execution of functions.

In this respect, the control concept described in [33] - and converted into pilot solutions - can be considered as most promising. Here, a CNC is regarded as a logical system, the software structuring of which is carried out by means of a two-level concept. It includes:

- The action level, which combines the set of actions as sequential operations of the subsystems of the control device; and

- The co-ordination level, in which single statements concerning the problem-solving and the parallel co-ordination of all actions are accumulated.

The modelling of the logic system CNC is carried out with the aid of logic expressions in the form of "when-begin" statements in one set of rules. They are stored as instructions and constitute the basic knowledge for solving various problems (execution of actions). Extensions or modifications, i.e. the increase of knowledge, are realized using novel approaches - firstly, by an externally organized teaching process and secondly by the automatic accumulation of knowledge based on experience.

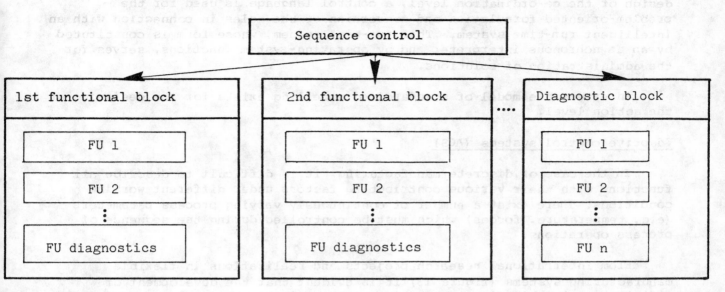

Figure 11. Software structure for MPST control [38]

FU = Functional unit

Figure 12. Structure of MPST control [38]

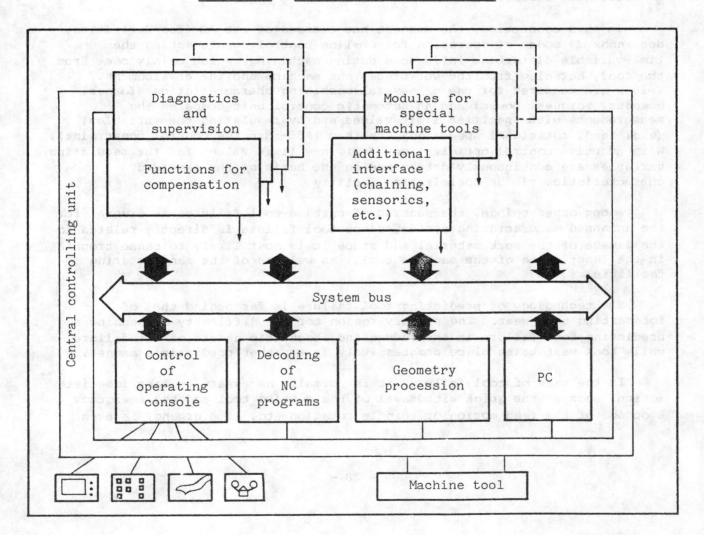

For the practical conversion of the control concept, especially for the design of the co-ordination level, a control language is used for the problem-oriented formulation and conversion of the rules in connection with an intelligent run-time system. This run-time system, whose form is constituted by an asynchronous interpreter and by operating-system functions, serves for the administration of solutions.

An extensible model of standardized modelling exists for the designing of the action level.

Adaptive control systems (ACS)

In the case of discrete manufacturing, it is difficult to describe all functions with their various contributing factors under different working conditions. There exist a number of continuously varying process parameters (e.g. temperature, forces) which must be controlled during the sequence of process operations.

From international research projects and realizations in flexible manufacturing systems (figure 13) it is evident that the development of efficient tool-monitoring systems forms the basis of the development of process-accompanying system components. The tool is the most critical element in the machining process, since it guarantees the undisturbed sequence of manufacturing operations.

In this connection, the concept has emerged of the AC (for definition, see annex I) modes of operation for on-line control counteracting the time-variable disturbing influences during machining, which mainly come from the tool, but also from the workpiece, the machine and the environment. Values are measured for one or several machining characteristics (forces, moments, torques, power) and an automatic control unit compares the measurements with specified limit values and by regulating the variables, (e.g. feed, rotational speed) adjusts them (ACC-adaptive control constraint). With adaptive control optimization (ACO), new limit values for the regulating variables are continuously determined on the basis of the recorded characteristics via an optimization facility.

Among other things, the sensing of cutting-tool failures is crucial for the unmanned manufacturing system, since tool failure is directly related to the damage of the work material and since it is most likely to cause trouble in the functioning of the machine tools, as well as of the manufacturing facilities [34].

The technology of predicting tool failure is far behind that of forecasting tool wear. The primary reason for the difficulty of sensing and predicting tool failure is the sudden and stochastic nature of the failure, while tool wear takes place progressively in a rather predictable manner.

In the case of tool failure, it is normally necessary to take immediate action, such as the quick withdrawal of the cutting tool and the emergency stoppage of the feed motion or spindle rotation etc. The urgency is even

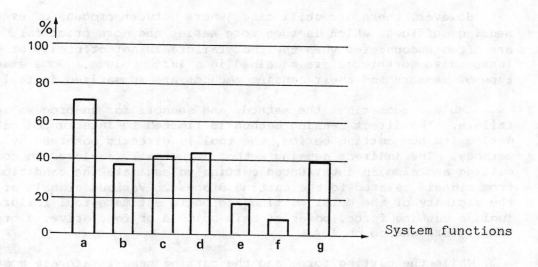

Figure 13. Frequency distribution of system functions carried out in flexible manufacturing [38]

a — Tool supervision
b — Measuring of the workpiece
c — Tool handling
d — Handling of the workpiece
e — Handling of the clamping device
f — Auxiliary functions
g — Error detection

Note: Analysis based on information drawn from engineering periodicals describing 146 FMC/FMS (or parts of such systems) in the years 1979-1985.

- 29 -

greater in the event of breakage or catastrophic failure of the tool. Thus in-process sensing is essential in order to avoid damage of the workpiece or the manufacturing facilities and to take quick remedial action.

However, there are still cases where between-process or even post-process sensing suffices, which is much more easier and more practical. Such cases are often encountered when the tool failure is not critical, or when inexpensive workpieces are machined in a large volume. Some examples of this type of sensors and their sensing methods are summarized in table 2 [34].

Table 3 summarizes the methods and sensors for in-process sensing of tool failure. The direct sensing method is limited to interrupted cutting, where, during its non-cutting period, the tool is directly observed by optical methods. The indirect sensing method is normally employed in continuous cutting and also in interrupted cutting to estimate the condition of the tool from signals related to the cutting process. Various signals are sensed in the vicinity of the metal cutting region to estimate tool failure; these include cutting force, power of main spindle or feed drive, vibration, sound temperature and so on [34].

While the cutting force and the cutting temperature are most sensitive to tool failure, the cutting temperature measurement is not much used at present, since it is troublesome and not very practical to employ on the shop floor. The correlations between changes in cutting force and tool failures of various kinds are well known, and most of the tool failure sensors in practical use or under investigation are based on some kind of cutting-force measurements. Typical sensing methods and sensors of the cutting force so far developed are summarized in table 4.

Table 2. Practical examples of between-process and post-process sensors of tool failure [34]

	Sensing method	Sensor
Direct sensing	Sensing of position or existence of tool edge	Optical sensor including fibre optics
	Sensing of tool edge position	Contact sensor
Indirect sensing	Measurement of dimensions and roughness of workpiece	Touch trigger probe, Analogue touch sensor, Non-contact proximity sensor, Optical sensor, Pneumatic sensor

Figure 14 shows the typical behaviour of forces after the occurrence of tool failure.

Table 3. Methods and sensors for in-process sensing of tool failure [34]

	Sensing method	Transducer signal	Sensors
Direct sensing	Direct measurement of failure	Light	ITV, Photonic sensor
Indirect sensing	Estimate from variation, increment and differential coefficient etc. of cutting forces	Cutting force	Strain gauge, force transducer, tool dynamometer
	Estimate from change and pattern etc. of cutting power or feed power	Power	Current meter, watt meter
	Estimate from level and spectra etc. of acceleration	Vibration	Piezo-electric acceleration pick-up
	Estimate from level of AE signal	Sound or AE	AE sensor
	Estimate from rapid increase of cutting temperature	Cutting temperature	Tool-work thermocouple
	Estimate from contact electric resistance between tool and work	Electric resistance	Volt meter
	Estimate from change of surface roughness	Surface roughness	Lamp and diode detector

Table 4. Typical sensing methods and sensors of cutting force [34]

	Sensing method	Sensor	Remarks
Direct sensing	Sensing by strain gauges (octagonal ring, or diaphragm-type sensor etc.)	Strain gauge	Dynamometer used mainly for research and development
	Sensing by combined strain gauge and acceleration pick-up	Strain gauge and ACC pick-up	
	Sensing by piezo-electric force sensor	Piezo-electric transducer	
	Sensing by load washer installed beneath tool post or table	Strain gauge, semi-conductor load transducer	Installed within machine tool
	Sensing by strain gauge attached to spindle	Strain gauge	
Indirect sensing	Estimate from deflection of spindle by proximeter pick-up	Eddy current type proximeter, capacitive type proximeter	
	Estimate from deflection of tool post by electric micrometer	Differential transformer	
	Estimate from power consumption of driving motor	Current meter, watt meter	Simplified sensing
	Estimate from hydaulic pressure of hydrostatic bearing	Pressure transducer	
	Estimate from slip rate of driving motor	Tacho-meter, slip meter	

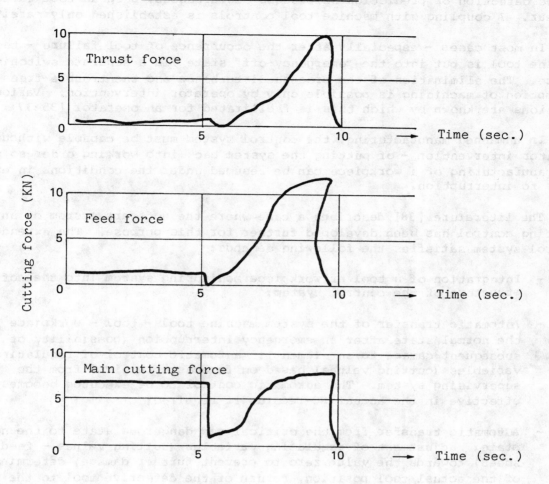

Of primary importance in the development of the above monitoring systems is the detection or prediction of process disturbances, such as tool failure or wear. A coupling with machine tool controls is established only rarely.

In most cases - especially after the occurrence of tool failure - the machine tool is put into the "emergency-off" state via a separate switching device. The elimination of the process disturbance and the trouble-free resumption of machining is possible only by operator intervention. Various solutions are known by which this is facilitated for an operator [35-37].

In unmanned manufacturing, the control system must be capable without operator intervention - of putting the system back into working order so that the manufacturing of a workpiece can be resumed under the conditions in effect prior to interruption.

The literature [38] describes a case where the software system of an existing control has been developed further for this purpose. The extended control system satisfies the following demands:

- Integration of a tool or workpiece monitoring system into the software structure of the control system;

- Automatic transfer of the system machine tool - tool - workpiece into the normal state after an emergency interruption (possibility of subsequent damage to an element): automatic control of regulating variables (cutting values) based on forecasts received from the supervising system. The automatic control to be executed becomes effective in the actual NC-record, if required;

- Automatic transfer from the critical and dangerous state to the normal state: adjustment of regulating variables (cutting values - feed, speed) towards the value zero to prevent further damage, determination of the actual tool position, return of the defective tool to the starting position, replacement of the tool and/or the workpiece, repositioning and resumption of the interrupted manufacturing process; and

- Execution of the functions of the information available in the control system.

Direct numerical control (DNC)

Direct numerical control (DNC) is a system used for several manufacturing, processing and measuring machines and for workpiece and tool manipulation and transport and storage systems which are equipped with digital control systems by a computer (see also definitions in annex I). The digital computer is suitable for the storage, collection and distribution of instructions concerning sequence of actions, paths and process state from and/or for the binary control systems. Communication can be performed bi-directionally in competitive operation according to programmed priorities between the control systems and the digital computer.

DNCs are characterized by the following general features:

(a) The comparative importance of the DNC basic function is declining;

(b) Since real-time data supply of NC-machining stations has ceased, the priority of the basic function over the extension function no longer exists;

(c) The computer-internal representation of NC programs in the form of NC control programs is decreasing in favour of more general forms of representation as source programs;

(d) The extension functions are increasing in number and influencing the basic functions;

(e) The interpretation of the extension functions is more and more becoming the starting point for the generation of the basic functions; and

(f) The degree of consistent information processing is continuously increasing, aimed at multiple use of already generated data sets.

The development of DNC was initiated to make possible the control of machine tools from a central computer. The idea was to take advantage of the time-sharing capability of digital computers, to replace the paper-carried information input in the rough workshop climate of NC- machines, and to make program handling less cumbersome. A characteristic feature of these systems is the time- and format-adequate distribution to the machine tools of the programs stored in the computer.

The general structure of DNC-systems comprises a central computer, one or several coupling elements and several NC-machines with an attached DNC-system. Various forms of execution and arrangements of the transfer channels are used between the individual units of equipment. The direct computerized guidance of NC-machines includes basic and extension functions. Whereas the fulfilment of the basic function was the purpose of the first DNC-systems, the extension function is already available as a result of the first steps in the further development of DNC.

For the broad effect of the extension function, unidirectional data transfer from the digital computer to the NC-machines was insufficient. There was need for bi-directional information transfer between the devices of the DNC basic structure with format- and time-adequate data distribution via a communication software module or bi-directional hardware equipment capable of communicating. Since the early 1970s, the availability of DNC-systems with this functional content has given rise to prospects of the development for workshop-wide process automation and automated, almost unattended manufacture. One more step towards bi-directional paperless information transfer between technical preparation and the workshop has been taken by adapting the hardware and software structure of DNC-systems to the advanced developments of NC-programming methods. This demands a variety of forms of communication with the DNC-computer across interfaces S1, S2 and S3.

The data transfer is classified as:

- S1, for communication with the NC-machines;

- S2, for communication with programming systems; and

- S3, for communication with different functional complexes for the control of systems and manufacturing.

The data-transfer mode at interface S1 is determined by DNC-machines. It is connected in real time for NC-machines and characterized only by the communication interface; it is free of real time for CNC-machines. The possibility of storing machining programs in CNC control units permits the transfer of whole NC programs. Hence, the supply of information does not have to keep up with the manufacturing process and may show a basic difference to the real-time program supply. This means that the software does not need to handle real-time interruptions and makes it possible to include functions with routines unfavourable to interruptions via interfaces S1, S2 and S3. Thus, it is possible to maintain a continuous flow of information on preparatory tasks and for the monitoring of actions which are closely related to processes. This provides a decisive pre-condition for the implementation of computerized programming.

The programming of NC programs for NC-machines of the DNC-system can be executed via interface S2. Dialogue programming on the display screen for the generation of source programs is possible without any restrictions. In the DNC-computer, the data are converted into control programs for NC-machines by the programming software and postprocessors for data transfer via interface S1. Interface S3 permits the integration of new equipment adequate for the increasing complexity of functions in flexible manufacturing systems. DNC does not serve in the supplying of these units of equipment with programs, but in the recording of system state quantities and the derivation of control functions for devices which are closely related to the machining process, the NC-machines and the transfer of information to superior control computers of factories.

The following developments for DNC may be expected:

(a) Complete utilization of CAD/CAM systems, which are characterized by paperless information transfer, from design and technology up to machining on CNC-machines by the use of coupling modules and CLDATA-type computer-internal sets of information;

(b) A workshop communication system for production process control by the interconnection of system-state information of the total manufacturing process and computer-aided instructions for planning and guidance; and

(c) Integration of process-accompanying monitoring systems, as well as of transport and storage equipment, to raise the degree of automated, almost unattended manufacturing, leading to a high degree of self-containment in flexible manufacturing systems.

Even the real-time guided NC-machines show essential qualitative gains. Together with advances in sensorics, informatics, and the development of equipment, these trends are the driving forces in the development of DNC.

II.1.2. Flexible manufacturing cells (FMC)

A manufacturing cell is a small grouping of machine tools or technological units which includes integrated subsystems that operate largely automatically and perform the following functions:

- To supply, handle and process workpieces and/or to join assembly units;

- To store and to change tools;

- To perform quality control; and

- To control the whole system under consideration of economy and security aspects (see also definition in annex I).

Workpieces of a single or several types can be manufactured to the end or the intermediate form in one or more set ups.

The flexible manufacturing cell (FMC) is best suited to the manufacture of families of parts of high variety and low volume. It can have a number of configurations, but generally it comprises more than one machine tool with some form of pallet changing equipment, such as an industrial robot or other specialized material handling device. In most cases, the grouping of machines is small and utilizes a common pallet or part fixturing device for the specific part requirements. Generally, the FMC applies a fixed process and parts flow sequentially between operations. The cell lacks central computer control with real-time routing, load balancing and production scheduling logic.

The equipment of modern FMC is increasingly automated. The extensive range of functions, the high precision in processing, the reduced time required for manufacture, the variety of products and virtually complete machining are fundamental factors giving rise to increased productivity and quality. Correspondingly, the demands to be met in flexibility of controls are very high.

Following the trend for flexible manufacturing cells and complete processing, manufacturing and handling processes are increasingly combined in one machine tool. The scope of functions of CNC controls which some years ago was limited to such fundamental operations as the interpolation of two or several axes and the response to some auxiliary functions has been considerably extended. These new functions include:

- Coupling of the machine tools/manipulation system;

- Internal automatic measurement;

- Coupling of external measuring stations with machine tools;

- Tool supervision;

- Distribution of control data for additional drives (e.g. autonomously driven tools in turning machines); and

- Co-ordination of several tool supports [40, 41, 42].

A new aspect in cost reduction is the application of CNC controls. The programming systems developed for this type of computer may be directly applied to the machine. Unmanned manufacturing secured during determined periods is increasingly realized. Thus, the use of adaptive control systems (ACS) increasingly gains in significance in:

- Providing for quick reaction to events which could cause subsequent damage to tools or working pieces (e.g. in case of tool breakage); and

- Recognizing conditions which could endanger the quality and the operational sequence, such as tool wear or chipping.

The AC systems are extensively based upon limit controls - Adaptive Control Constraint (ACC), i.e. a process determining characteristic is led to a preset limit value during processing. These characteristics are registered by sensors. By this AC, in addition to the protective function mentioned above, the basic machine time is reduced and the tools and driving power are used more extensively [43, 44] (see chapter II.1.1).

Before the FMC is put into operation, the control of the manufacturing sequence must be set. Owing to the high degree of automation and the high level of integration of handling systems, it is very difficult to look into the working zone of the machine tool. In the past, program tests were generally carried out on the machine tool itself, but such tests cannot be performed in highly automated manufacturing plants. Consequently, graphical simulation is becoming of increasing importance.

In turning operations, there are already systems allowing complete testing of programs on the screen and workpieces can be manufactured by automatic operation. In milling operations, however, owing to the unavailability of reasonably priced 3-D geometric modellers, testing of programs still requires the presence of the workpiece. Furthermore, the working and clamping conditions of the workpiece must also be taken into account.

Simulation system requirements

In designing a dynamic simulation system, all the characteristics of the manufacturing process must be included in the model. The algorithms of the CNC must be extensively tested in order to simulate an NC program in real time.

In contrast to turning, in milling the shapes of the clamping devices and workpieces are varied and more complex and it is therefore insufficient to use a definite geometry file. To enlarge the file and to facilitate the description of difficult workpiece contours, a comfortable editor with graphic support is required.

Figure 15 gives a survey of requirements for dynamic simulation systems.

Another area which should be mentioned is the simulation of robots. As robots are mainly programmed by the teach-in method, the elaboration of the handling program is of great importance (see chapter I.1.4).

II.1.3 Flexible manufacturing systems (FMS)

A flexible manufacturing system (FMS) includes at least three elements: a number of work stations, an automated material-handling system and system supervisory computer control. The FMS is typically designed to work for long periods with little or no operator attendance. FMS meets the demand for machining in a batch environment where equipment dedicated to high volume production, such as transfer lines, is cost prohibitive. FMS, unlike the transfer line, can react quickly to product and design changes (see also definition in annex I).

Automatic tool changing, in-process inspection, parts washing, automated storage and retrieval systems (AS/RS) and other computer-aided manufacturing (CAM) technologies are often included in FMS. Central computer control over real-time routing, load balancing and production scheduling distinguish FMS from FMC. Flexibility is the most important aspect of FMS [39, 45].

Communication between individual systems is of special importance. The CNC control as a link to the machine tool has in this case to provide for appropriate interfaces and its scope of functions must be appropriately enlarged.

To interlink and synchronize the individual activities of an FMS requires flexible and precise production planning and supervision of manufacturing; the density of the data to be processed for the individual components is very high (figure 16).

Simulation models are used for analyses which could not be carried out on the machine itself for both time and cost reasons. With the help of computer simulation, it is possible to establish the optimum layout of technological equipment before actually modelling it and of technological sequences before implementing them. In technical applications, there is need to distinguish between structural simulation in planning time and process simulation in real time (figure 17).

The purpose of structural simulation is productivity analysis for flexible manufacturing systems with a view to determining the most effective form of manufacture. This is done by drafting various manufacturing systems

Figure 15. Demands for dynamic simulation system [43]

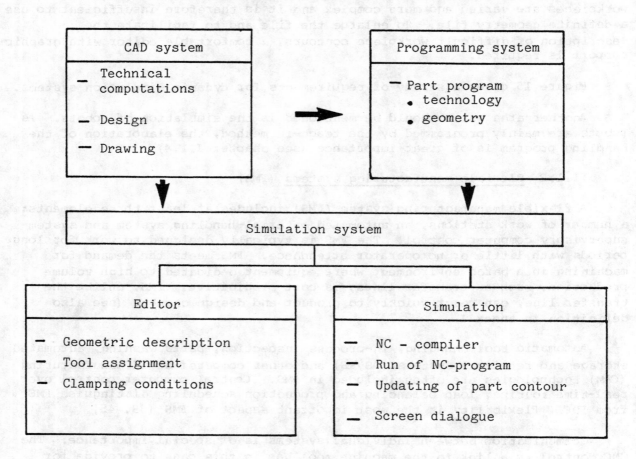

- 40 -

Figure 16. FMS part programmer [49]

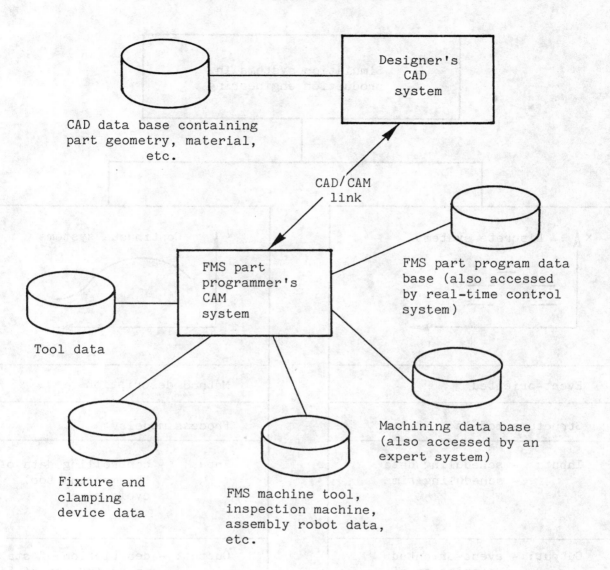

CAD data base containing
part geometry, material,
etc.

Designer's
CAD
system

CAD/CAM
link

FMS part
programmer's
CAM
system

Tool data

FMS part program data
base (also accessed
by real-time control
system)

Machining data base
(also accessed by an
expert system)

Fixture and
clamping
device data

FMS machine tool,
inspection machine,
assembly robot data,
etc.

Figure 17. Application of simulation in production engineering [43]

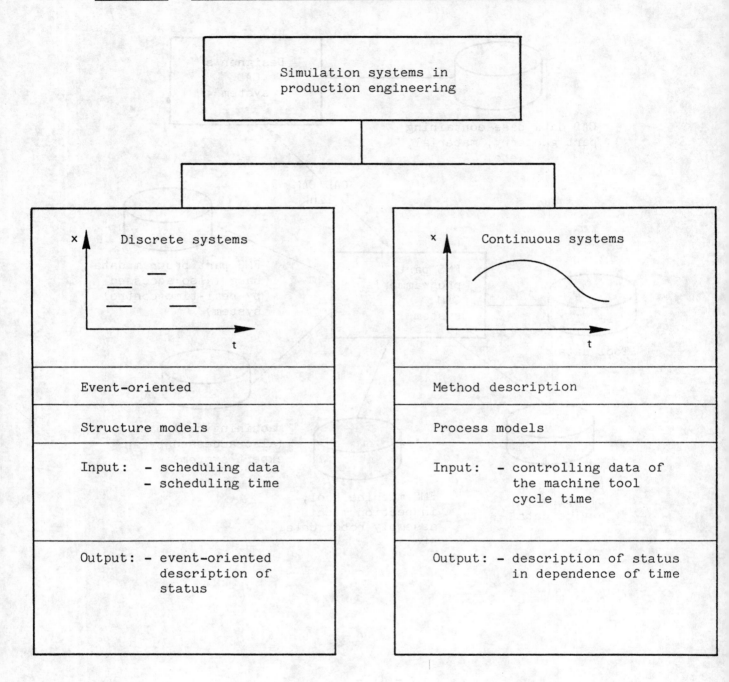

on the basis of relevant planning data and the simulation of their behaviour in relation to the given manufacturing task and the fixed concomitant conditions.

Dynamic (continuously in time) simulation is focused on the direct elaboration of NC-programs and the updating of the representation of sequences of operations. The exact sequence of these operations from the raw material to the final part can be observed on the graphical display. This method requires the imitation of essential algorithms of a CNC in the simulation computer. Its main advantage lies in the fact that it facilitates the identification of programming and sequential errors before the workpiece is actually set up for machining. [43]

The further development and utilization of FMS necessitates the standardization of the software making up the control system. A typical control system comprises three stages hierarchically working together:

- Operative production planning;

- <u>Ad hoc</u> control and supervision of manufacturing;

- Control of the technological equipment.

Further information on this topic is provided in annexes II-IV.

A standard architecture of software for FMS proposed by the USSR is contained in annex V.

II.1.4 Industrial robots

II.1.4.1 General remarks

International agreement has as yet not been reached on the definition of the term "industrial robot". For the purpose of uniformity, ECE has adopted the definition proposed by the International Organization for Standardization [1].

The industrial robot is an automatic position-controlled reprogrammable, multi-functional manipulator having several degrees of freedom capable of handling materials, parts, tools, or specialized devices through variable programmed motions for the performance of a variety of tasks. It often has the appearance of one or several arms ending in a wrist. Its control unit uses a memorizing device and sometimes it can use sensing and adaptation appliances that take account of environment and circumstances. These multi-purpose machines are generally designed to carry out repetitive functions and can be adapted to other functions without permanent alteration of the equipment.

II.1.4.2. Programming of industrial robots [1]

The gradual extension of robots into new application areas, namely in small-batch production, brings with it a new problem - the effective programming of the robots. While robots were used mostly in mass production, the question of their programming was not as vital. A manipulation program was developed, debugged, verified and then used for a very long period, sometimes for the whole life of the robot. The employment of robots in small-batch production, in flexible manufacturing cells and in other similar applications, however, requires frequent changes in manipulation programs calling for the creation of special data files for use in connection with changes in the manufacturing process or its environment.

Effective programming methods for industrial robots had therefore to be developed and applied. The rapid development of control systems for industrial robots will facilitate the development of improved robot programming methods. Control units are usually built on the basis of microprocessors (Intel 8080, 8086, Motorola 6800, 68000 and others). These microprocessors, with a word length of 8 or 16 bits, permit employment of various types of problem-oriented languages, special high-level languages or assembly languages.

The following basic types of industrial robot programming methods are currently available (see figure 18):

Programming at the machine

(a) Manual programming

 - Hard-wired control

Usually supplied by the robot manufacturer, it is an unalterable control system designed for specific operations by a single-purpose robot. [2] This way of programming is employed, for instance, for material manipulation, where high precision of manipulation operations and changes in manipulation programs are not required.

 - Point-to-point (PTP) control

PTP control is used for applications with a small number of points in individual axes. Programming is divided into two parts:

 (i) Programming of position, realized by the adjustment of mechanical stops in individual steps and axes; and

---

[1] Based on a contribution by the Government of Czechoslovakia.

[2] Not an industrial robot in the narrow sense of the definition (see above), but rather a manipulator.

Figure 18. Programming methods for IR [23]

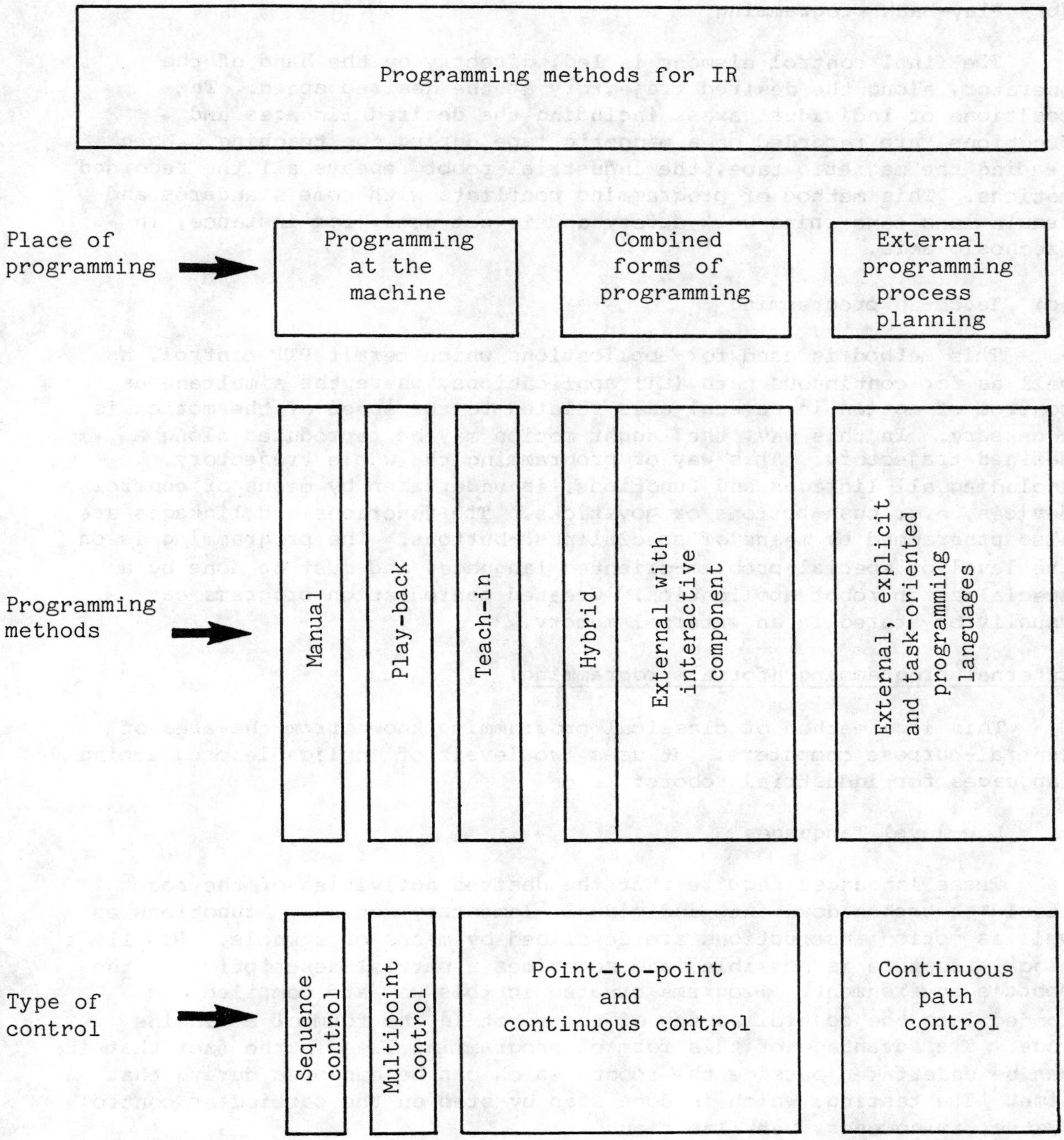

Programming methods for IR

Place of programming ➜	Programming at the machine	Combined forms of programming	External programming process planning

Programming methods ➜	Manual	Play-back	Teach-in	Hybrid	External with interactive component	External, explicit and task-oriented programming languages

Type of control ➜	Sequence control	Multipoint control	Point-to-point and continuous control	Continuous path control

- 45 -

(ii)   Programming of sequences and linkages, realized by means of programming matrices or sequential automats.

(b)   Play-back programming

The final control element is led, directly by the hand of the operator, along the desired trajectory at the desired speed. The positions of individual axes, including the desired linkages and functions, are recorded on a magnetic tape during the teaching. When reading the magnetic tape, the industrial robot repeats all the recorded motions. This method of programming conflicts with some standards and regulations concerning work safety and is not used, for instance, in Czechoslovakia.

(c)   Teach-in programming

This method is used for applications which permit PTP control, as well as for continuous path (CP) applications, where the simultaneous control of motion in several axes related to the speed of the motion is necessary. In this way, the taught motion may be reproduced along a defined trajectory. This way of programming the whole trajectory, including all linkages and functions, is undertaken by means of control devices, e.g. push-buttons or joysticks. The functions and linkages are also programmed by means of special push-buttons. The programming is on the level of special problem-oriented language, and must be done by a specialist in robot application. Created manipulation programs can usually be stored in an external memory.

External programming (formal programming)

This is a method of classical programming known from the area of general-purpose computers. It uses two levels of applicable programming languages for industrial robots:

-   Low-level languages

These languages require that the desired activities of the robot should be broken down into individual elementary motions. Functions as well as motion instructions are described by means of symbols. Usually a block structure is possible, and sometimes a partial description of the robot's environment. Programs created in this way are compiled and loaded into the control system of the robot in the form of a machine code. The advantage of this form of programming lies in the fact that it can be undertaken outside the robot, which can be employed during that time. The testing, which is done step by step on the particular control system, is however, very laborious.

-   High-level languages

The use of these techniques is at present still rather unusual for the programming of robots. However, the majority of robot manufacturers are working on this problem. These programming languages are for the most part constructed as special high-level problem-oriented languages. They provide a complex description of the robot's activities as well as of its environment, and possibly also of dynamic changes of the environment. The programming is sometimes undertaken in a dialogue mode; the description of the environment and its dynamic changes can be registered visually, by means of a camera. Such programming systems permit very effective programming of robots, but the development of the necessary systems software is very demanding. Increasing demands for complex functions of industrial robots will, however, make the creation of such programming systems unavoidable.

The methods of robot programming described above are realized by means of programming tools of various levels of sophistication. Usually, certain programming languages allow for several types of robot programming methods.

- Microprocessor low-level languages

These languages are used in the lowest level of robot programming, with the exception of hard-wired programs or programming by means of mechanical stops or programming matrices. The commands are aimed at the conversion of position co-ordinates into adequate control of drives. Analogically, the basic functions are programmed, such as clamping of gripper etc. This way of programming usually permits only PTP control. Commands are created for a certain type of robot control unit and are incompatible with other systems.

- Assembly languages

These languages are usually used in the teach-in method and possibly in the play-back method. It is also possible to form manipulation programs in a language close to the assembly language, known from the area of general-purpose computers. Contemporary robots software is on the level of well-elaborated assembly languages and serves to describe the complete trajectory of the robot, including specifications of all accessible functions. A manipulation program is often constructed in a dialogue mode, with the symbolic instruction recording mode being employed to define motion.

The motion can be specified in the PTP or in the CP way. The programming languages of this class permit branching of the program, conditional commands and loop definition. Subroutines can usually be employed. The software supports the editing of users' programs.

Some of these languages also include high-level commands, such as the command for transformation of co-ordinates or circle interpolation.

- High-level languages

  . Structured programming level

The use of programming languages of this class make possible the complex motion control of the robot and its functions. In most cases it is possible to apply transformation to describe the robot environment and define the status of variables. Parallel programming and a high degree of environment interaction is provided.

These languages replace programming of the PTP type and operation description with more complex motion definition. The possibility of describing the environment of the robot's working area is also provided, as well as the modelling of conflictless operations. A considerable advantage is the possibility of modelling outside the robot's control system, i.e. "off-line" creation of the manipulation programs. The need for a skilled programmer, acquainted with the area of the robot's application, may be considered a disadvantage.

  . Special problem-oriented languages

The aim of these programming languages is to reduce the burden of programming and to make the most effective use of the advanced features of high-level languages. These types of programming languages are as yet only in the early stages of development, and not in practical use.

The development of special problem-oriented languages is an area which is attracting considerable attention and which represents the advanced trend in the programming of robots. But their high level requires environment-description modelling, artificial intelligence methods for decision-making and interactive debugging systems. It is evident that the development of programming languages of this class is very demanding on time and financial resources. In spite of all these disadvantages, however, sophisticated industrial robot applications will necessitate programming systems of that class.

II.1.4.3. Trends and developments in IR application software

For the typical application fields of industrial robots - handling, assembly and machining - three fundamental fields of further development should be emphasized. These are: robot performance, robot sensory feedback, robot systems interconnection, and robot kinematics.

In this context it is interesting to note that, for general requirements for robot capabilities in nearly all fields of application, standardized off-line programming is utilized.

Users require a range of standardized programming languages at each level of robot application, including languages that allow better interfacing with other computer-integrated manufacturing equipment [51].

As concerns further development of robot technology, such as

- Greater speed and acceleration,

- Improved positioning accuracy,

- Improved repeatability and

- Improved reliability,

these are mainly linked with specific demands for the appropriate control software.

In the planning area a number of requirements are evident. An overall requirement in this context is for the development of a systematic approach and software tools to support the manufacturing engineer in planning and designing a robot-based manufacturing system.

The overall technological development is increasingly characterized by the fact that various types of industrial robots will be used as fundamental components of complex automation. Thus it is necessary for all planning tasks to be put on a uniform level in order to permit development and use of a high level of industrial-robot programming using CAD/CAM systems and CAD data banks. This requirement can be achieved by the use of off-line programming.

As with the NC programming of machine tools, programming of industrial robots attains a higher level with the aid of an external computer. The essential advantages of external programming methods are:

- Rational program generation through the utilization of high-capacity computer techniques;

- Reduction of the unproductive idle time of industrial robots; and

- Possibility of integration into efficient CAD/CAM systems and interconnection with suitable data banks.

The basic idea in programming industrial robots is that few statements are made in advance in the external programming, while all detailing of the sequences of motions is carried out with computer assistance in accordance with the specific character of the industrial robot and the problems of manipulation encountered. A general classification of programming languages for industrial robots is given in figure 19.

II.1.5 Automated material-handling equipment

Material handling (see also annex I), includes all the activities associated with the movement, handling, packaging and storage of substances in any form [53]. Automated guided vehicles for material handling and transportation, automated workshop storage systems, storage and integrated inventory control are gradually gaining acceptance in leading industrial companies.

<u>Figure 19.</u>   <u>Programming languages for IR</u> [23]

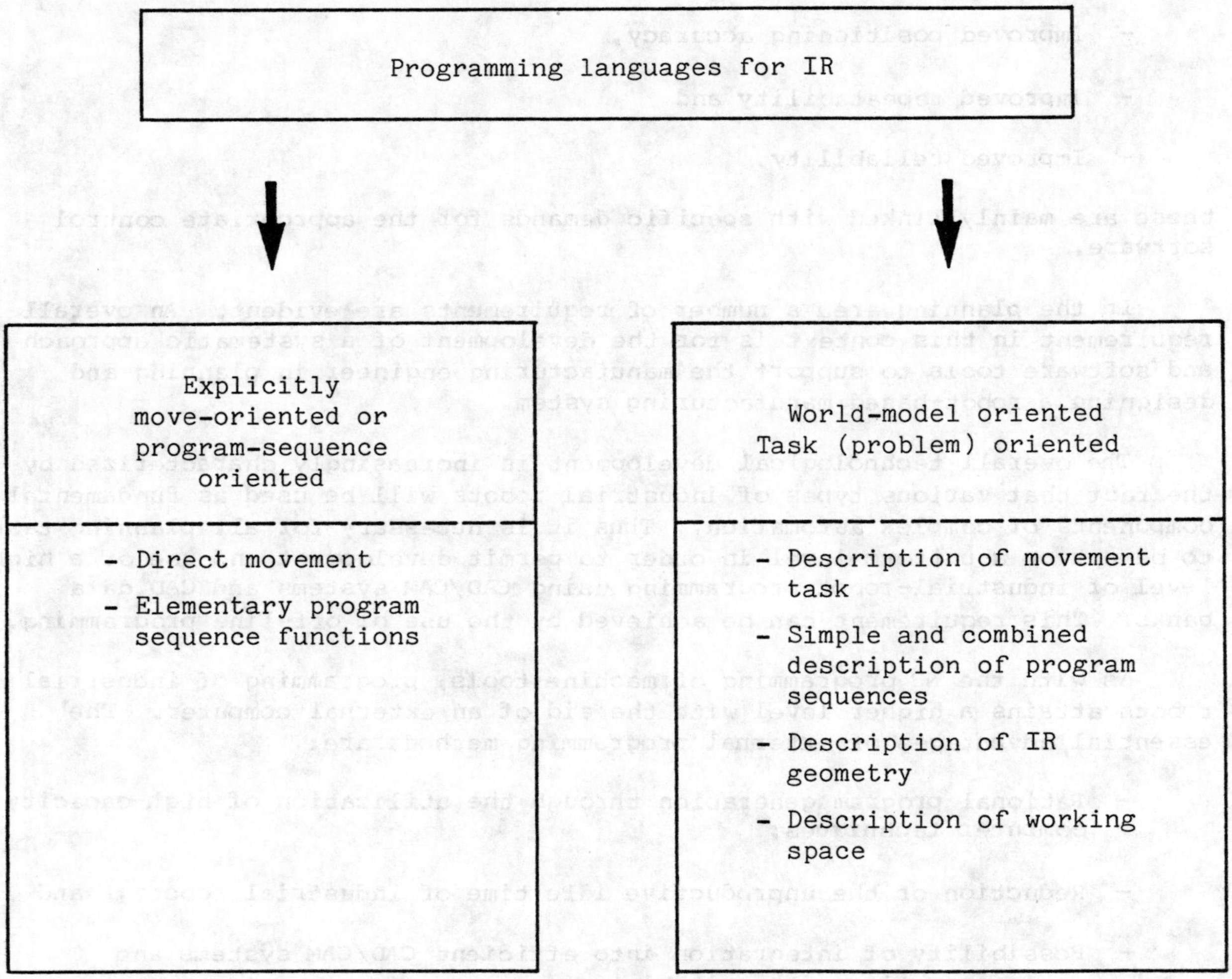

Programming languages for IR

Explicitly move-oriented or program-sequence oriented	World-model oriented Task (problem) oriented
- Direct movement - Elementary program sequence functions	- Description of movement task - Simple and combined description of program sequences - Description of IR geometry - Description of working space

Examples

VAL   (Unimation)   ROBEX   (Robot-EXAPT)
SIGLA   (Olivetti)   RAPT   (Robot-APT)
                              AUTOPASS (IBM)

The need for increased automation of the auxiliary process "material handling" is due to the following fact [53]. In a typical manufacturing enterprise, material-handling functions account for 25 per cent of total employees, 55 per cent of all factory space and 87 per cent of production time. A central part of material-handling systems are automated guided vehicles ((AGV) - for definition, see annex I). In a modern manufacturing enterprise, automated manufacture departments and flexible manufacturing systems are linked by automated guided vehicles.

Proceeding from the requirements of product and process planning and from the stochastically changing requirements of the manufacturing process, AGVs assure a continuous flow of parts, modules and products. In 1985, over 500 AGV systems were in use in Europe, involving some 5,000 vehicles [52]. The AGVs collect palletized loads from the incoming goods conveyors. Light barriers or other control equipment at the end of each conveyor are broken by a pallet, and these trigger a call for an AGV via the control computer.

The components of an AGV system are: [52]

- The vehicles themselves;

- The guidance and information transfer system, usually floor-based;

- The load-transfer equipment; and

- The vehicle- and traffic-control system.

The co-operation of these components is illustrated in figure 20.

The route control of an AGV system consists essentially in the transfer of data between the central controller (be it man or computer), the floor system and the vehicle. The data to be transmitted covers position information and commands which give rise to reaction both on the vehicle and in the floor system. The transfer methods most commonly in use are:

- Electromagnetic switches on the vehicle and read contacts in the floor system;

- Permanent magnets in the floor system and read switches on the vehicles;

- Separately controlled coils integrated in the guide wire and antennae on the vehicle;

- Infra-red transmitters and receivers;

- Radio receivers and transmitters; and

- An inductive data-transmission circuit with a transmission antenna on the vehicle.

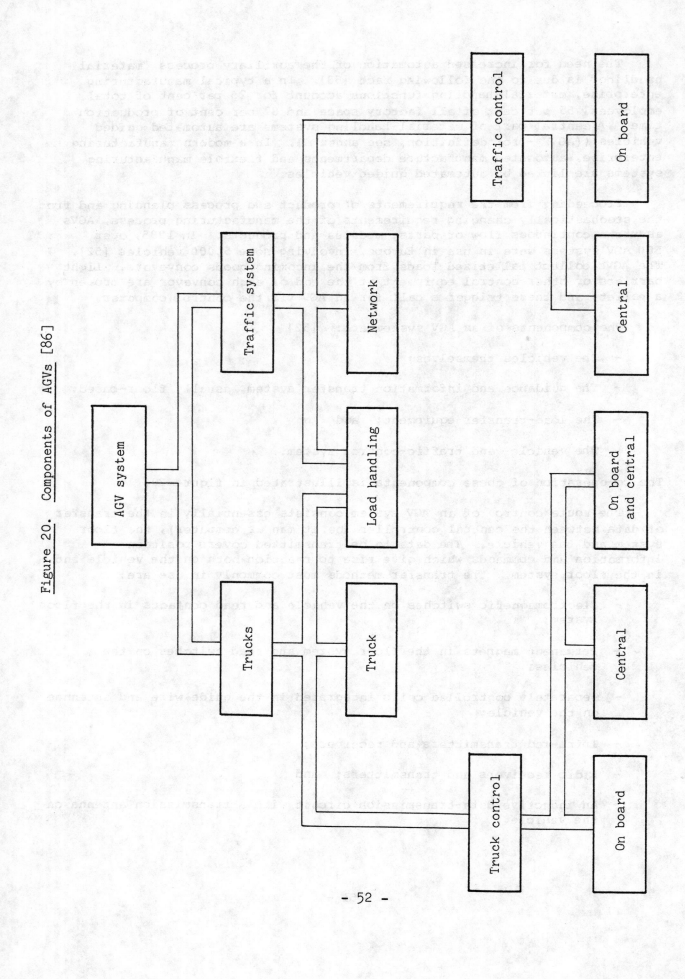

Figure 20. Components of AGVs [86]

- 52 -

Infra-red transmitters and receivers can transmit switching commands and transfer the destination address of a load to the vehicle. The information needed by the vehicle to reach its destination will have already been supplied by the operator and is stored in a route-program memory.

Radio links with the vehicles have the advantage of providing constant communication with the control unit. There may be problems, however, in achieving interference-free transmission.

For automatic storage, efficient hardware and software are required. To guarantee permanent and reliable readiness for operation, the use of synchronous duplex computers is recommended.

To control the sequences of loading and unloading functions, a special stock-keeping system is required which is integrated into the total manufacturing control. At the same time such a system of stock-keeping represents a link between the central mainframe computer and the controlling installations of the conveying and handling devices within the store.

The system of stock-keeping controls the following functions:

- Registration of the input order, distribution to palletized loads;

- Arrangement of storage places according to the nature of palletized loads;

- Observance of optimization criteria and criteria of safety;

- Registration of removed orders and insertion according to established priorities; and

- Taking of inventories.

II.1.6   Computer-aided design/Computer-aided manufacturing

II.1.6.1   General

The terms CAD and CAM (for definitions - see annex I) cover general information processing in the fields of:

- Computer-aided design;
- Computer-aided planning of manufacturing; and
- Automated manufacturing.

Whereas in the 1950s and 1960s a number of software packages were developed and used to support engineering decisions and routine tasks, the 1970s and 1980s saw the start of the transition to coupled and integrated systems.

On the basis of data available in the internal data base, these systems control all the principal tasks of production from the first general design stage up to quality checking and control of the final product. Today, a multitude of program packages are available (see figure 21) which, upon appropriate combination, should lead to real integrated solutions of automated design and manufacturing problems using CAD/CAM.

II.1.6.2 System characteristics

CAD/CAM systems are characterized by coupled or integrated processing of geometric, technical, technological, economic and organizational information. CAD/CAM solutions differ from so-called "island solutions" by the possibility which they offer for integration.

The application range of the main components of CAD/CAM systems is described in figure 22. The structure depicted in the figure is largely based on relevant Japanese sources. The following main aspects should be noted:

-    The field of CAD comprises design, modelling and documentation activities related to parts, tools, fixtures, appliances, special machines, technological processes, relevant production plant facilities, etc. An overlapping of some CAD functions with some functions of CAM (in particular computer-aided processing (CAP)) can be noted;

-    The term CAM always includes technological planning and manufacturing and manufacturing control and supervision activities; and

-    The special terms CAP and CAPP (computer-aided process and production planning) refer only to technological planning.

While this broad definition of CAD/CAM covers nearly all the engineering activities in an enterprise, certain limits should be taken into consideration.

Actual CAD/CAM systems ready for use are always designed for specific fields of application. Examples of such applications are:

-    Design and manufacture of printed circuit boards;

-    Design, production planning and manufacture of rotationally symmetric components;

-    Design, production planning and manufacture of prismatic parts; and

-    Design, production planning and manufacture of parts with sculptured surfaces.

Components of a CAD/CAM system may in some applications be used individually.

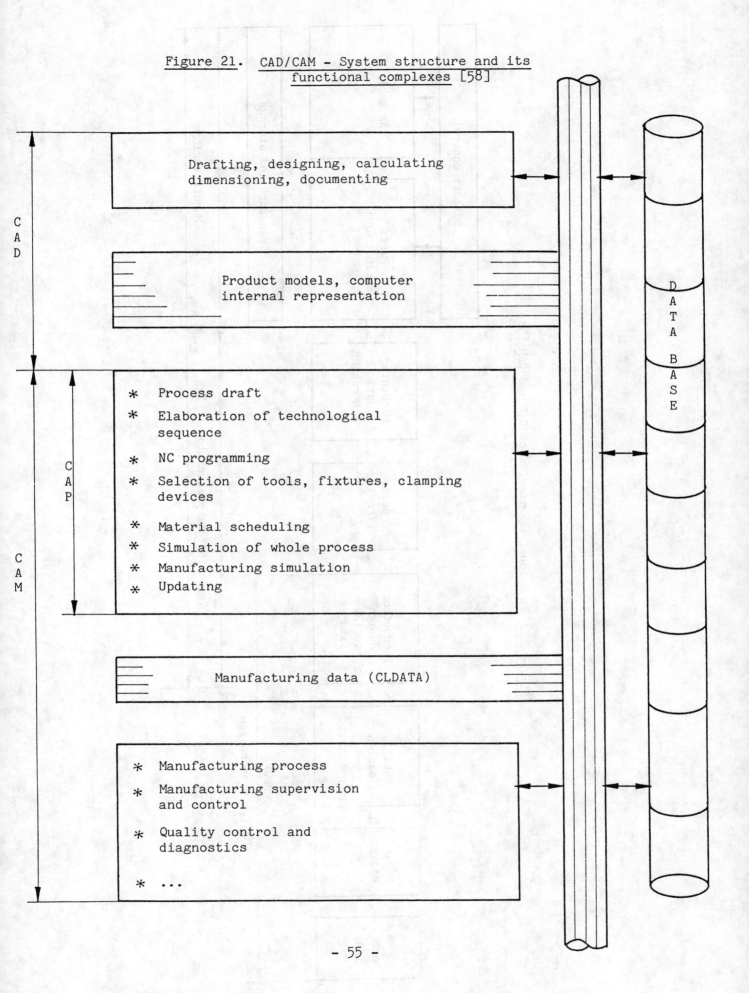

Figure 21. CAD/CAM – System structure and its functional complexes [58]

Drafting, designing, calculating dimensioning, documenting

Product models, computer internal representation

* Process draft
* Elaboration of technological sequence
* NC programming
* Selection of tools, fixtures, clamping devices
* Material scheduling
* Simulation of whole process
* Manufacturing simulation
* Updating

Manufacturing data (CLDATA)

* Manufacturing process
* Manufacturing supervision and control
* Quality control and diagnostics
* ...

CAD

CAP

CAM

DATA BASE

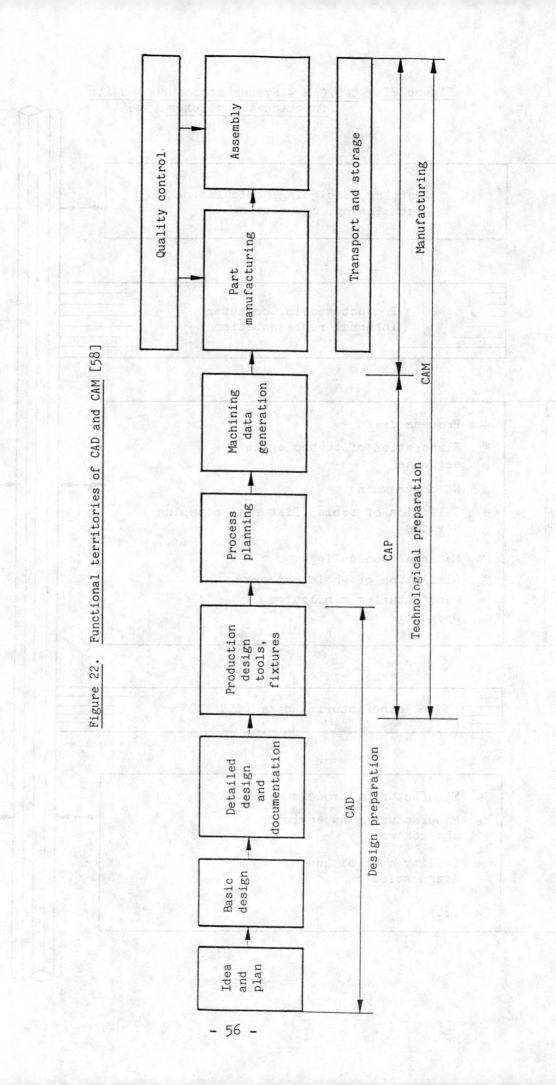

Figure 22.   Functional territories of CAD and CAM [58]

When creating CAD/CAM solutions, one should bear in mind that

- The level of automation varies between the fields of activities and within the fields themselves (i.e. technical design planning, conventional and automated manufacturing, process organization and preparation, etc.);

- The dynamics of technical development have a varying effect within those fields of activities; and

- Sometimes differences exist in standards and norms used within enterprises, combines and companies.

The following are amongst the principal features of general high-level CAD/CAM systems:

- Preference for the use of interactive modes of operation for different classes of activities, on the basis of specialized work-stations;

- Availability and utilization of extensive program and data banks at various access facilities and application of supporting systems guaranteeing portability;

- Uniformly designed interfaces between program components;

- Facilities to apply graphic data processing;

- Application of program elements for problem-oriented input and output (including problem-oriented input languages);

- Application of uniform software design concepts;

- Modular designed software; and

- Ability to expand the systems.

II.1.6.3 System components

Generally, CAD/CAM systems consist of the following system components (for definitions, see also annex I).

System component "Design" (CAD)

The "design" component is responsible for all aspects related to the design and drafting of a technical object (e.g. of a machine, a unit or a component). The result is the external graphic display or internal computer description of the object in a form ready to be used for manufacturing. In design, the development of a product passes through the following phases:

- Determination of the problem;
- Elaboration of principal solutions; and
- Design (drafting dimensioning, specifications, documentation).

- 57 -

In addition, the designing department must perform the following tasks:

-   Evaluation of tests of functional models or manufacturing models; and

-   Preparation for manufacturing (including provision for lists of parts on the basis of updating).

Software for CAD

As regards systems software, the general world trend is towards systems using the various forms of UNIX which permit simultaneous interactive execution of extensive tasks largely focused on 32 bit mini- or microcomputers in networks. Programming languages are FORTRAN 77, C language, etc. Hardware independence is ensured by means of graphic standards GKS, CORE and the standard for product modelling. On the level of application software for CAD, the trend is towards three-dimensional systems which permit, by means of the volume-elements method, the execution of calculation and plotting tasks as well as automation of plane and special geometric tasks. The software is based on the finite element method.

Two-dimensional systems used in application areas of engineering permit the creation of drawings of both primitive and parametric objects. They work with broken lines, points, circles and ellipses and provision can be made to include the text material, attributes, marks, etc. related to the objects. With these systems it is possible to solve plane geometric tasks, work with text, cross-hatching and dimensioning, work with libraries and do archiving. Transformations, such as shift, rotation and affinity, are also included in the systems.

Handling or treatment of general surfaces and curves is considered a separate problem. Sophisticated CAD systems are able to solve this problem in plane (2-D) as well as in space (3-D).

An important component of CAD software is its ability effectively to handle data-base management. In this regard, many complex software systems are being developed on the basis of users' own CAD software. Such systems tend to be specialized for certain product areas. The systems have the advantage of being compact, but the disadvantage of incompatibility and narrow specialization.

System component "Production planning" (CAP or CAPP)

The "product planning" component includes all the steps required for the elaboration of a manufacturing sequence for a product. The result is the external graphic display or internal computer description of the entire manufacturing process. Applications of the production planning component include, in particular, the following:

-   Planning of the sequences in the manufacturing process;

-   Material-requirement planning and scheduling;

-   Scheduling of manufacturing equipment; and

-   Updating.

Software for CAP or CAPP

The CAP or CAPP systems employ common basic software and basic graphic software, which is supplied as a standard with mini- and microcomputers. Besides that, standard data-base management software is also available.

The applications of CAPP software can be divided into three principal areas:

- Programming systems for automation of production-process design. These systems work either on the basis of type or group technology or on the principle of multilevel synthesis from elementary standard elements of technological process. Over 50 such systems have been developed all over the world. Recently, the development of these systems has tended to create and employ interactive systems, i.e. systems allowing dialogue between the designer of the production process and the computer. In the case of modern systems, the dialogue is carried out with the help of computer graphics.

- Programming systems designed for the programming of NC-machines. These use languages which can be integrated in CAPP, and which are extensively employed for the technological rationalization of the production process. These modern systems make extensive use of computer graphics to facilitate dialogue between the NC-machine programmer and the system. This makes it possible to speed up the work and to simulate production processes.

- Application program packages for automated design and technological preparation of the production of special manufacturing facilities, where practically all types of tools - cutting tools (shaped tools and cutters, pull broaches), forming and casting tools, clamping tools and gauges - are designed by computer. This case also employs graphic dialogue between the specialist and the system.

In recent years, the trend has been to develop systems for application areas, such as those referred to above, not as independent program units, but rather as parts of complex systems for the automation of technical and other activities of the enterprise.

System component "Manufacturing" (CAM)

The "manufacturing" component includes all activities required to carry out and to supervise the manufacturing process. The input variable comprises the complete description of the manufacturing process. The result is the finished part. Fields of application of the system component include:

- Automated manufacturing (increasingly resulting in fewer but more complex operations), including manufacturing control;

- Handling and removal of waste; and

- Supervision.

The development and use of automated manufacturing with fewer and more complex operations is accompanied by a considerable increase in expenses for applications software for sensors, handling devices, industrial robots and automatic diagnostics.

Software for CAM

The current dominant trend in the development of CAM software is the gradual realization of integrated systems based on the concept of computer-integrated manufacturing (CIM) (see chapter II.1.7). This results from increased technical control over CAD/CAM systems, their rapid growth and the need to integrate them into the overall management and information processing structure of industrial enterprises.

The software required for the management of production and technological processes linked with CAPP as well as the planning and administrative activities in an enterprise is of a modular and hierarchical structure. Principles of distributed intelligence, high-level programming languages and standardization in the area of communication protocols in computer networks architecture are consistently applied. Non-standard situations are solved by human intervention.

Efforts are currently being directed towards improving the adaptability of software to various computer and user environments. This will lead to the elimination of hardware-dependent features of the software and to the possibility of generating applications software for particular user conditions. Great emphasis is placed on software for acquisition, transmission and distribution of technological and operating data and on the structure and generation of data-bank systems, for both centralized and distributed data-base concepts.

Programming systems with artificial intelligence elements are gradually being implemented. The related software is created with the help of software-engineering tools and methods.

The operation of CAM systems calls for two types of primary data:

-   Construction and technological data (for example, data on process planning, fixtures, tools, testing equipment, NC-information, etc.). Based on these data, order-independent manufacturing documents are compiled, using production management data (i.e. batch quantity, probable production period etc.).

-   Data on production organization (data concerning manufacturing orders, material requirements data, etc.). These data are elaborated on the basis of construction and technological data to provide order forms at the due date, specifications on material and manufacturing equipment at the proper time, as well as other documents required for planning and the disposition of parts.

The efficiency of automated solutions is determined by the completeness, topicality and quality of both types of primary data. The elaboration, administration and further processing of both types of primary data should be an integrated part of implementation strategy.

II.1.6.3 Hardware and software requirements

II.1.6.3.1. Hardware

To implement integrated automated systems, various computer-hardware solutions are possible:

(a) Computing equipment located at the work-place of the designer, production planner and manufacturing engineer, etc.;

(b) Computing equipment located in EDP centres with terminals at the work-places of the designers, production planners and manufacturing engineers;

(c) Computing equipment in hierarchical nets (local area networks (LAN)); and

(d) Computer networks.

In production planning, specialized work-stations are in widespread use; they are efficient and simple to use with peripheral devices with an alphanumerical or graphic display device as their principal input/output component.

In general, the following computer system hardware components are available:

- The input and editing station to input geometric and other information. Usually, it includes a digitizer – this is in most cases designed as a panel consisting of an alphanumerical keyboard.

- The design and editing station usually comprising a graphic and an alphanumerical display unit. By applying these two devices in production planning, an interactive mode can be used to solve many problems.

- The central processing unit. Depending on the size of the installation, this normally has a range of from at least 64 K to 256 K Byte – in some cases even more – internal memory capacity (16- or 32- bit instructions with direct, indirect or indexed addressing) and in general can be linked with several input work stations. In addition to processing the programs which are transferred to it, the central processing unit controls data transfer between the central computers and all peripheral units.

- External storage devices comprising mainly magnetic discs and tapes.

- Output devices. In addition to the commonly used output devices, such as tape punchers and fast printers, plotting equipment for graphic output is frequently included in basic configurations.

II.1.6.3.2. Software

Currently the development of software lags behind that of hardware. Therefore, the development of both systems and applications software and especially of software tools is of crucial importance for the future utilization of CAD/CAM

applications. According to [60] for the achievement of appropriate CAD/CAM
solutions, the following criteria govern CAD/CAM system selection:

- Adaptability to operational - minimizing customer investment
  conditions   for installation, commissioning
                                   and development

- User friendliness   - usable for drafting and design
                                   applications, capable of fast response
                                   and with potential for future development

- Reliability   - proven hardware and software and a
                                   minimum maintenance requirement

- Productivity   - the system must have been demonstrated
                                   in service on similar work

- CAD/CAM capability   - the facility to use common data at all
                                   stages of the manufacturing sequence -
                                   from design through to manufacturing

- Transparency   - capable of being operated by engineers
                                   without special computer experience

- Competitive price   - economic viability both with respect to
                                   the cost of development and operation
                                   and to the overall return on investment

A uniform software architecture could improve the efficiency of the system.
One possible four-level architecture is illustrated in figure 23.

In addition it should be kept in mind that the applications software could be
improved by a modular structure. Figure 22 depicts principal complexes of typical
CAD/CAM modules. As is illustrated in figure 24, certain modules are required for
the comprehensive classification of tasks into operational informational structures
and others for specific classes of tasks (e.g. 2-D and 3-D modelling).

The total organization and interaction of the individual application
components requires appropriate CAD/CAM frame systems.

In order to facilitate adaptation of CAD/CAM systems to different operational
requirements and for their further development, functional openness is of
particular importance. In this respect, the following requirements are important:

- Concerning the manufacturing process: Coupling possibilities to the
  computer-aided control of manufacturing and its supervision should be
  envisaged.

Figure 23. General software architecture for CAD/CAM [58]

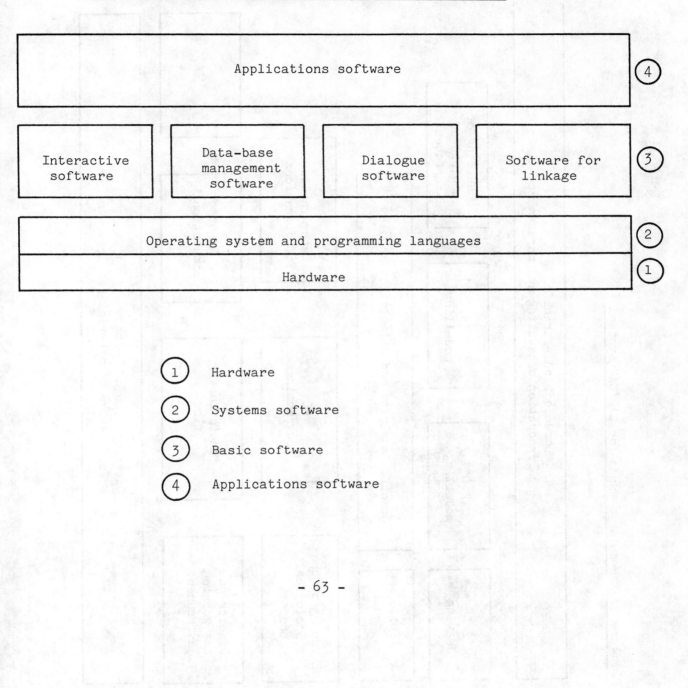

Figure 24. Modular demands for CAD/CAM systems [58]

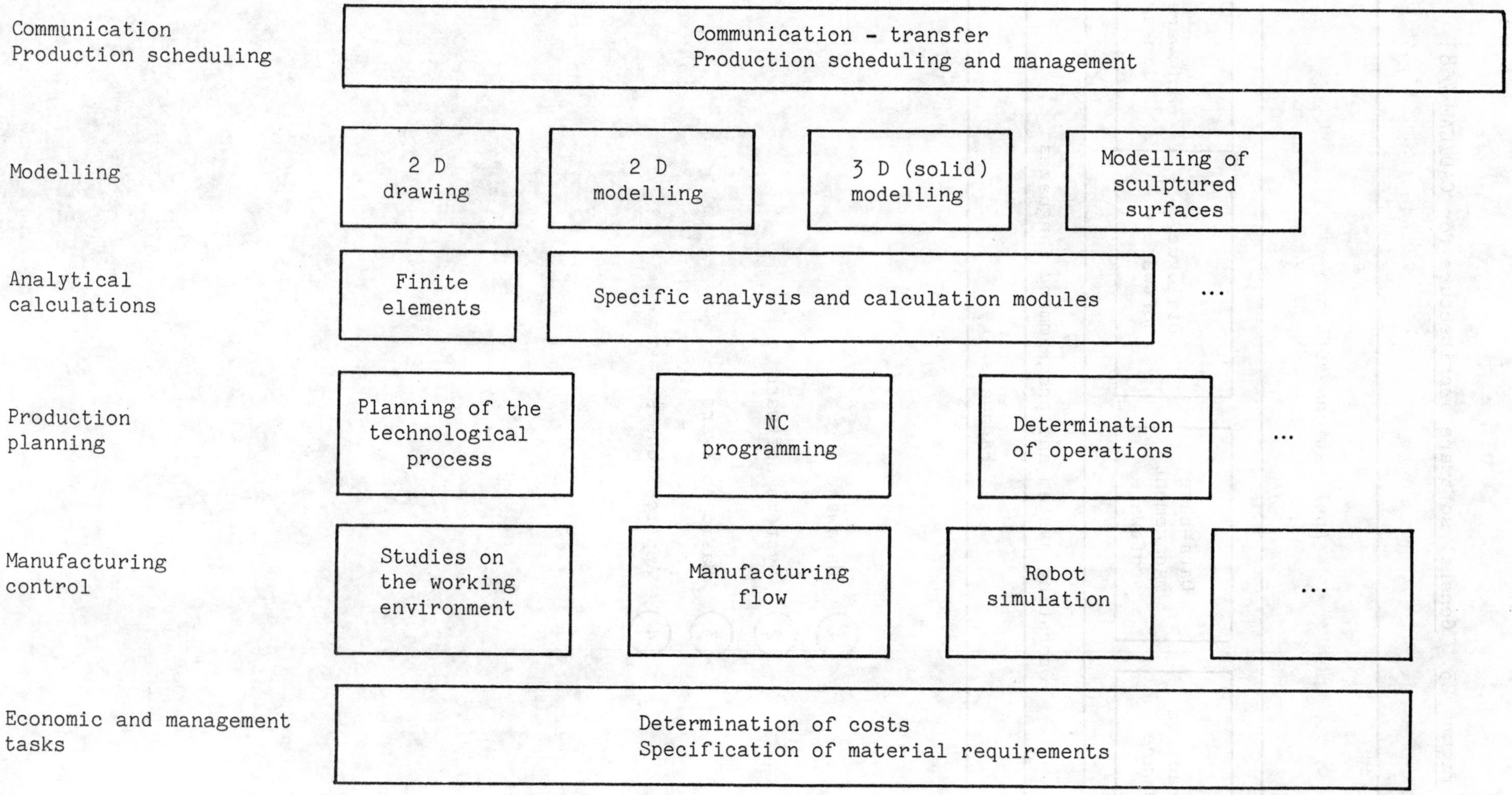

Communication
Production scheduling

Communication – transfer
Production scheduling and management

Modelling

2 D drawing	2 D modelling	3 D (solid) modelling	Modelling of sculptured surfaces

Analytical calculations

Finite elements	Specific analysis and calculation modules	...

Production planning

Planning of the technological process	NC programming	Determination of operations	...

Manufacturing control

Studies on the working environment	Manufacturing flow	Robot simulation	

Economic and management tasks

Determination of costs
Specification of material requirements

- Concerning the user: The user interface should give possibilities of interaction techniques which can be influenced by the user; i.e. specific forms, by which the user can modify and adapt, according to his requirements, the screen layout and the dialogue type, control and content.

- Concerning advances in manufacturing equipment: The possibilities of enlarging the system should be taken into consideration, beginning with alphanumerical up to graphical displays, from two-dimensional to three-dimensional models.

The software characteristics and criteria described above are important for the further development of CIM (see chapter II.1.7).

II.1.6.4. Data banks as central elements in CAD/CAM systems

As illustrated in figure 21, the data bank is a central element within a CAD/CAM system. Geometric, technological and organizational data used by different modules to solve given problems must be systematically classified and controlled. Concerning this extensive software complex associated with CAD/CAM systems - for which there is still no internationally standardized solution or fundamental structure - some principal functions are described below [23].

The data bank and its associated operating software controls the long-term recording and management of all data. These data pertain to design and technological planning tasks and results. The data are divided into general information, design information, drawing information, technological planning information and NC data. The processing programs for data recording and data management are provided by the data file system. If the information is provided in separate individual data files, the data file management is generally accomplished via the operating system of the computer. Apart from data recording and management - functions under the control of a central data bank - data handling can be performed by a data file distributed to several logically interlinked data archives.

Computer-internal representation (CIR) includes all current object data and all relations between object data which are necessary for an object image and object processing in the functional modules. The features of the CIR are the data structure, the interface with the computer-internal representation, and the model algorithms used. The data structure is the logically set up correlation of data by which the object is described. The range of application depends on the layout of the CIR. The more precise and complete the object image is developed, the wider the application spectrum. The layout and design of the CIR are determined by the process-technologies and user-specific requirements.

Data transfer between the functional modules of the CAD/CAM system is performed via the system interfaces. The result is the logic interlinking of communication modules and the data file. Consequently, these system interfaces are of particular importance for integration measures.

The efficiency of data bank systems and the associated support systems is enhanced by manifold and different user interfaces. Important user interfaces in connection with essential functional complexes for data handling are described in table 5.

Table 5. User interfaces and their functional complexes

Command and access variation

- New commands

- Users identification

- Access authorization

Variation of data model

- Modification of external model

- Definition of screen recordings

- Optional printings

Secondary data check

- Test of check digits

- Test of plausibility

- Determination of default declaration

Secondary data delivery

- Determination and transfer to linked modules

- Access to other data bases

Insertion of new modules

- Problem modules

- Modules of inquiries

II.1.6.5. Internationally applied interfaces in CAD/CAM systems

Interconnection between the software components of a CAD/CAM system requires appropriate interfaces (for definition, see annex I). These should be internationally standardized. Of special importance are those interfaces which are common to many systems, irrespective of their field of application. The standardization of such interfaces would contribute considerably to the rational development of software, to software exchange and to software portability. A distinction can generally be made between hardware and software interfaces. Within the scope of the present study, only software interfaces are considered.

Software interfaces are possible between

- The components of a CAD/CAM system;

- Different CAD/CAM systems; and

- The CAD/CAM system and its environment.

II.1.6.5.1 Product model

The functioning of a CAD/CAM system is based on the exchange of data between the components of the system and a central data bank and between different CAD/CAM systems. The simplest way to solve the interlinking of the components and systems is to create an adaptive system for each interface. However, it should be noted that, in practice, this is difficult. For example, when connecting four systems, six adaptive systems are required and, when adding a further system, an additional four adaptive systems are needed. Thus it is evident that it would be almost impossible to provide such adaptive systems for very large systems.

Efforts to elaborate standardized interfaces to solve these problems started in September 1979 with the development of a first version of IGES (Initial Graphics Exchange Specification) within the ICAM project (Integrated Computer Aided Manufacturing) of the United States Air Force. The various stages of the development of this standard are outlined in chapter II.3. It is expected that a first version of an internationally accepted standard will be available for industrial application by 1990.

II.1.6.5.2. Graphics interface

In view of the wide variety of graphics application programs and of graphics devices available, there is a need to standardize the associated software. Furthermore, the user functions associated with the input and output of graphics-related data must be hardware independent.

The Graphical Kernel System (GKS) provides an interface between the application program and the configuration of graphics input and output devices. It includes all the functions required for interactive and passive graphics routines.

The hardware independence realized in GKS permits the free exchange of user programs between different GKS installations. Thus, instead of programming the different graphics devices, the user simply refers to standardized graphics (program) functions. The advantages of the system are as follows:

- Transfer from one graphics device to another without any need for adaptation;

- Standardization of application programming;

- Reduction in programming and training costs;

- Simplified introduction of new graphics devices without the need to develop new software, provided specific modules exist; and

- Interchangeability of graphics user programs.

At present there are only a few CAD/CAM systems which offer all these advantages. However, it should be stressed, that optimum application and user efficiency depend on meeting the above demands. In addition, the still more complex informational structures which will be elaborated in the future (see chapter II.1.7) will necessitate the realization of a consistently high level of CAD/CAM solutions.

II.1.7 Computer-integrated manufacturing (CIM)

The speed of technical innovation in industry is accompanied by a parallel development of various levels of automation. Computer-aided techniques are at present still in the early stages of implementation in factories. While more work is needed before the key technology of CAD/CAM can become a part of complex integrated systems, other more complex kinds of integrated information processing will need to be developed for future use. The term CIM is not only employed at conferences and in literature, but suppliers of computers and machine tools offer software and automation solutions entitled "CIM", using the term as a synonym for maximum system integration. This trend was particularly evident at the EMO '85 exhibition at Hannover in the Federal Republic of Germany.

Although there is as yet no unambiguous definition of CIM as there is of CAD/CAM, it should be emphasized that CIM characterizes a new aspect of the application of information processing in industry. Two proposals for a definition of CIM may be found in annex I.

II.1.7.1 CIM - structures and functional complexes

In the case of CAD/CAM, there is a need for integration of design and production planning functions and partly or fully automated manufacturing processes. Advances towards CIM could be characterized by:

- A higher level of integration through the interconnection of all activities within the factory environment;

- A higher degree of adaptability and flexibility of the system, with provisions for feedback at all stages and sequences of decision-making; and

- Further development of knowledge-based systems in several fields and kinds of application (see chapter II.1.8 and figure 27).

Systems interconnection

The following are some of the more important activities in the factory environment whose interconnection in a powerful communication network constitutes a prerequisite for CIM (see figure 25):

- Assortment scheduling (determining demand, long-term disposition, capacity balancing);

- Order scheduling (inventory control, material-flow control), and production scheduling and management (product specification, material and goods handling, scheduling of capacities, management of ordinary data);

- Computer-aided production planning and manufacturing;

- Manufacturing control and supervision (short-term scheduling, documentation, order tracking, staff and order disposition);

- Registering of status information on, for example, manning tables, orders, manufacturing costs, material flow, tools, fixtures, clamping devices, handling devices, etc.;

- Generation and compilation of data for quality control; and

- Collection of sales data.

The accessing of common data bases by terminals and work stations in all departments of a factory and an appropriate integrated system architecture is a prerequisite for computer-aided scheduling, production planning, manufacturing and quality control, as well as for monitoring sales.

System feedback

In computer-integrated manufacturing, all steps in manufacturing and decision-making beginning at the stage of the receipt of orders are controlled by computers. In addition to the highly developed CAD/CAM systems characterized by uni-directional data flow from management via scheduling and production planning to manufacturing, future CIM structures will include various feedbacks. These will call for the integration of all computer-aided departments, with a view to quickly adapting the conditions and/or status of the factory, or, in other words, its overall input and output, to changing targets, i.e. optimizing the whole factory environment and controlling all steps in the hierarchical decision-making process in accordance with the actual target. Integration and internal possibilities of feedback are necessary prerequisites.

- 69 -

Figure 25. Interconnection of the main system elements
in CIM structures [58]

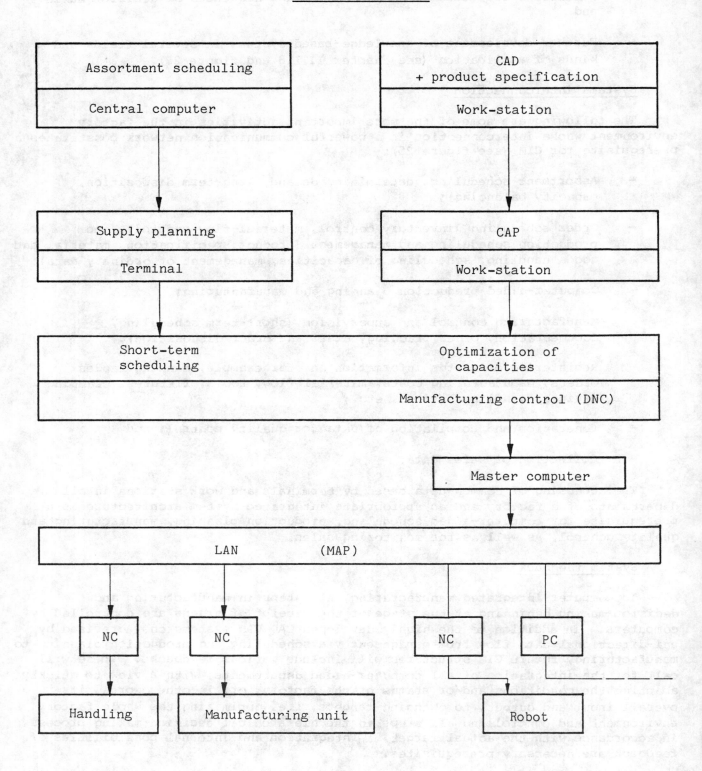

Computer-integrated manufacturing is still on the drawing board. At the time of preparation of this study no operationally functional systems of this kind are known to exist in industry.

The role of computers in CIM will not be limited to computation, dimensioning, drafting and documentation. Production planning and manufacturing will be undertaken by computer, using simulation methods. In the design stage, the information on applicable manufacturing methods, costs of manufacturing and the efficiency of manufacturing systems will be determined by the computer system. This will permit the economic estimate and comparison of the technological variants. Furthermore, the above-mentioned feedbacks influence other activities, such as integrated scheduling and production planning, integrated manufacturing and manufacturing control and integrated quality control.

In the above respect, the current developments in "off-line programming" should be noted. These will allow for the efficient and quick introduction and adaptation of many CIM features into existing manufacturing organization. Progressive computer systems and software packages which control a network of material handling, order management and registration of manufacturing data and which perform production planning tasks will form the central part of CIM structures.

CIM applications will result in:

-    Reduction of raw material and in-process stocks;

-    Reduction of production lead time;

-    Improved control over the fulfilment of contracts and meeting of deadlines; and

-    Increased flexibility in meeting market demands.

II.1.7.2. Communication requirements of CIM

The integrated data flow within CIM structures is based on a complex communication network covering all working departments of a factory. The data flow occurs in two main directions:

-    Centralized information in a top-down manner; and

-    Status information and information on production preparation and manufacturing in the reverse direction.

The hierarchic structure of an enterprise may not be limited to one single integrated communications bus, but a hierarchy of buses and communication devices may be necessary. An example of this approach is illustrated in figure 26.

Figure 26. Communication transfer in CIM structures

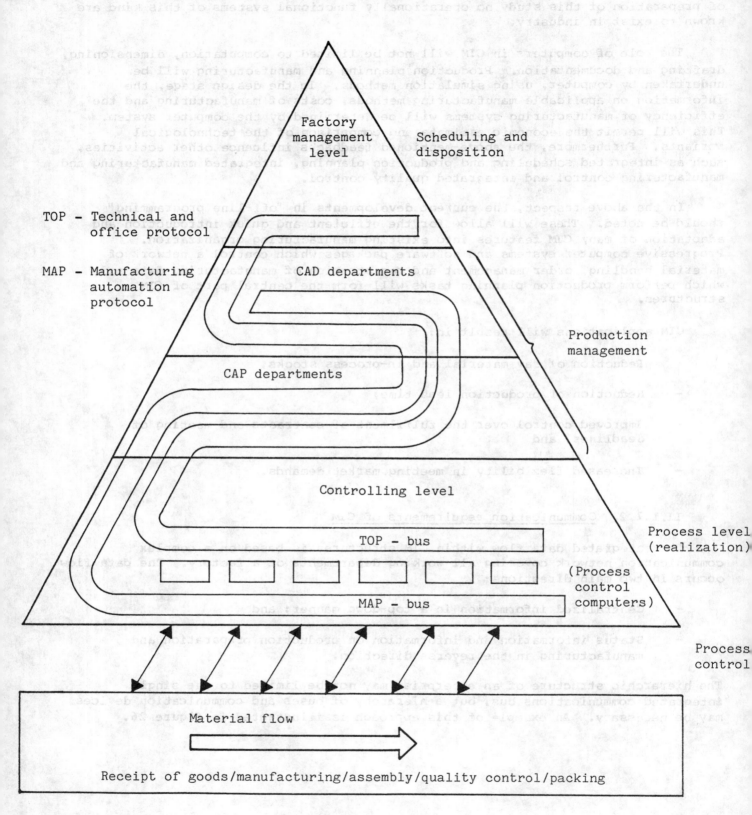

Factory
management
level

Scheduling and
disposition

TOP - Technical and
office protocol

MAP - Manufacturing
automation
protocol

CAD departments

Production
management

CAP departments

Controlling level

Process level
(realization)

TOP - bus

(Process
control
computers)

MAP - bus

Process
control

Material flow

Receipt of goods/manufacturing/assembly/quality control/packing

The above principle of control levels may be considered to be universal as applied to production data. Within a large factory, the principle is secured by a computer hierarchy assigning control computers with terminals to each level. For smaller factories, this may be carried out using only one computer which can supervise both the functional division of, and the assignments to the various control levels. Developments in this field are based on generalized universal strategies, although each system solution may call for a particular data flow system in conformity with individual product and manufacturing structures and requirements.

CIM structures may have to be adapted to individual factory conditions in a more comprehensive manner than CAD/CAM systems. Unambiguous interfaces and modular software components will be an essential prerequisite.

At the process level, speeds of data transfer of 10 Mbit/sec will be needed for real-time operation. At pre-processing stages, data-transfer speeds reaching 10 Kbit/sec will be sufficient for transmitting production and management data not critical in respect of time.

In CAD and CAP systems, large amounts of data have also to be processed at the speed of 10 Kbit/sec.

The structure of communication protocols defined by ISO - the ISO 7-layer model - provides the technical base for all types of communication networks.

Informational requirements of CIM systems - summary

Communication:	Possibility of communication in a network of systems from various suppliers
Distributed data processing:	Microprocessor-based work stations, master computer
Compatibility:	Applicability of computer hierarchies (families) covering a wide spectrum
Modularity:	Possible step-by-step development of hardware and software
Openness:	Existence of standardized interfaces to external hardware and software
User-orientation:	Unambiguous and user-oriented, convenient operations.

Effects gained through the realization of CIM structures

The implementation of CIM - as indicated by the term "integration" - is the application of computers in order to achieve complete integration of all manufacturing. Data flow is multidirectional - from materials purchasing to product design, development, manufacturing, quality assurance to warehousing, handling, sales and distribution. All staff, including senior management and machine operators, etc. are provided with the necessary information and all parts, tools and material etc. are supplied to the manufacturing equipment at the appropriate place and time by means of CIM.

II.1.7.3.   Trends towards Mechatronics

An interesting development in the direction of highly flexible automation structures is "mechatronic technology" [54]. "Mechatronics" is a popular term originating in Japan which essentially refers to the process whereby the technology of modern mechanical systems have been replaced by electronic controls. It represents an effective combination of hardware - both mechanical and electronic hardware - and software solutions in manufacturing engineering. A typical advantage of mechatronics technology results from the application of advanced electronics to mechanical control. The impact of mechatronics applications can be demonstrated by the following examples:

1.   A conventional mechanism is simplified by utilizing a microprocessor-based control system.

2.   The motion of the mechanism is controlled by appropriate software.

3.   A new function or high performance feature, which cannot be realized with the conventional mechanism, is implemented by software with no or few changes in the mechanism.

For example, a threading function of a lathe requires a gear-transmission mechanism by which the tool-feed motion can be synchronized with the main spindle rotation. However, a CNC-lathe has an electronic synchronization mechanism eliminating the need for a physical gear - transmission. This greatly simplifies the tool-feed drive mechanism and the machine set-up and gives greater versatility to the programmable threading operation. The microprocessor-based on-line calculation for synchronized motion allows the cutting of specially designed thread such as tapered thread or consecutively variable (increasing or decreasing) lead thread, which is almost impossible in the conventional or hard-wired NC-lathe. With the developments taking place in automated manufacturing, there is a critical demand for skilled technicians in the mechanical industries today. In order to meet the demand, mechanical engineering students should be trained in mechatronics technology, with the main emphasis on the application of microprocessors to machine control.

Generalized mechatronics control call for the following functions:

-    A programmable automatic control function

-    A manual control function

-    A preparatory function

-    A system diagnostic function.

Each function consists of several subfunction modules, as shown in figure 27. The function should become active when an appropriate operation is engaged. Typical operations required in mechatronics control are:

- Automatic operation
- Manual operation
- Preparatory opertion
- Emergency and initialization operation.

II.1.8   Artificial intelligence

II.1.8.1 General trends and basic concepts

There is as yet no uniform definition of the concept of artificial intelligence. According to Turing, one of the pioneers of computer technology, a machine behaves in an intelligent manner when the questioner is not able to determine whether he or she is confronted by a human being or an automaton [23]. Although this definition is plausible and to the point, it covers only the imitation of human intelligence. If one takes into account technical interrelationships, the following definitions seems to be more valid: "artificial intelligence is a category of highly automated information processing in technical systems" [23]. Other concepts are described in annex I.

Of great importance are the so-called expert systems (ES). Expert systems are designed for automated problem-solving in special applications. Taking the part of an expert in automation-aided consultations, they assist and advise the user in problem solving. The basis of knowledge which contains the relevant facts and rules is an essential characteristic of ES (see figure 28). Referring to the field of application - technical and engineering-technological problems - one speaks of mechanical or machine intelligence.

II.1.8.2 Trends towards the development of artificial intelligence

Work is being undertaken in many countries aimed at the development of highly efficient and intelligent problem solving systems termed "artificial intelligence". Although the term requires very careful consideration and explanation, three aspects could be discussed in the context of production:

- Methods of knowledge support leading towards expert systems may have to be used for external production planning;

- In manufacturing and supervision of manufacturing, new principles of machine intelligence (more complex software solutions, sensorics, diagnostics) may have to be applied; and

- Although there are CIM systems without artificial intelligence or expert systems, there are also CIM situations that would be hard to conceive without artificial intelligence.

As knowledge processing develops, it will become more important for automation. However, automation systems will always have need of data processing capacity, as its lack would make impossible the use of modern control algorithms, such as those used in multivariable adaptive control.

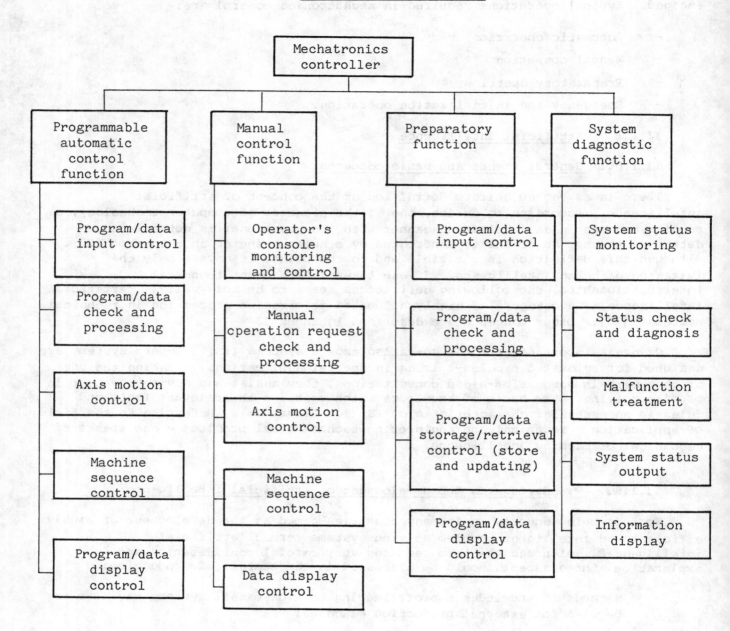

Figure 27. Functional structure of generalized mechatronics controller [54]

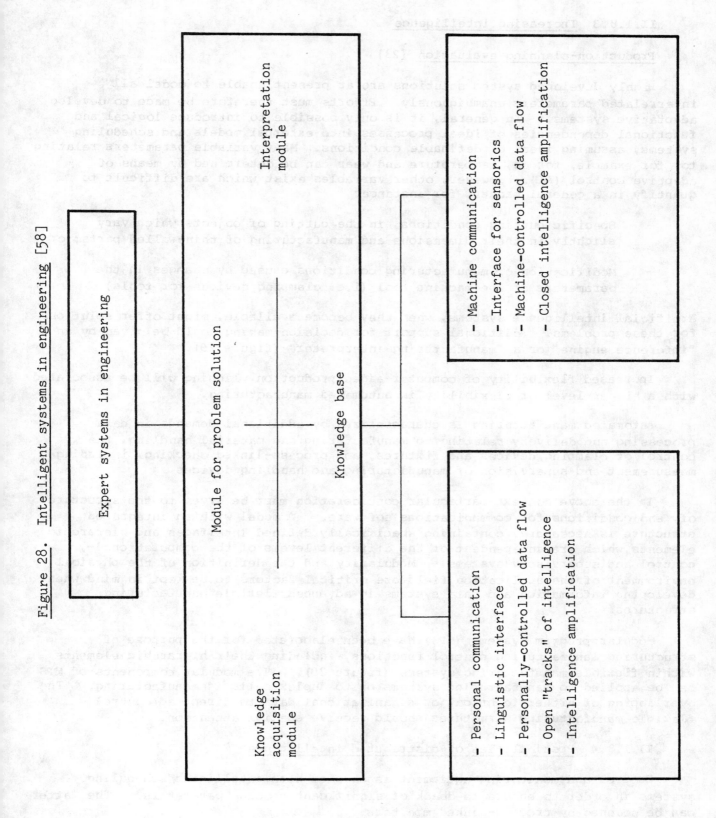

Figure 28. Intelligent systems in engineering [58]

Intelligent systems in engineering

Expert systems in engineering

Knowledge acquisition module

Module for problem solution

Interpretation module

Knowledge base

- Personal communication
- Linguistic interface
- Personally-controlled data flow
- Open "tracks" of intelligence
- Intelligence amplification

- Machine communication
- Interface for sensorics
- Machine-controlled data flow
- Closed intelligence amplification

II.1.8.3 <u>Increasing intelligence</u>

<u>Production-planning evaluation</u> [23]

Highly developed system solutions are at present unable to model all
interrelated parameters unambiguously. Efforts must therefore be made to develop
adaptative systems.   In general, it is only possible to introduce logical and
functional dependencies of ideal processes into external models and scheduling
systems, assuming ideal or definable conditions. Many variable parameters relating
to, for example, forces, temperature and wear can be determined by means of
adaptive control (AC). However, other variables exist which are difficult to
quantify in a general manner, for instance:

-       Specific cutting conditions, in the cutting of objects which vary
        slightly in their dimensions and manufacturing of thin-walled parts; or

-       Modification of manufacturing conditions caused by changes in the
        parameters of the machine tool (i.e. clamping devices for tools).

Artificial intelligence systems, when they become available, might offer solutions
for these problems. Additional support for decision-making could be given by an
"inference engine" or a "manufacturing interpreter" (figure 29).

Increased flexibility of computer-aided production planning will be associated
with a higher level of flexibility in automated manufacturing.

Automated manufacturing is characterized by additional demands in data
processing and delivery relating to manufacturing and material handling, the
control of clamping devices and fixtures, and process-linked checking, including
measurement and supervision of manufacturing and handling devices.

In the above context particular consideration must be given to the structure
of, and conditions for communications software. A model with an integrated
structure is necessary, containing specifically defined interfaces and hierarchic
elements which are independent of the different levels of the computational-,
control and supervisory systems. Modularity and the definition of the physical
environment of the application field are critical factors to be kept in mind when
developing information and data systems in advanced flexible manufacturing
structures.

Modular program systems (MPS) have been elaborated for the purpose of
structuring manufacturing control functions, including their hierarchic elements
within flexible manufacturing systems (figure 30).   The modular components of MPS
can be applied to manufacturing systems or to duplex cells in manufacturing.   The
overlapping of either technical or organizational data to direct ad control
flexible manufacturing structures should receive special attention.

II.1.8.4 <u>Possibilities of distributed intelligence</u>

Control and monitoring equipment is coupled with intelligent scheduling
systems in order to ensure feedback of significant process parameters.   The latter
can be scanned by process-linked monitors.

Figure 29.   Possibilities of artificial intelligence (inference
engine, manufacturing interpreter)

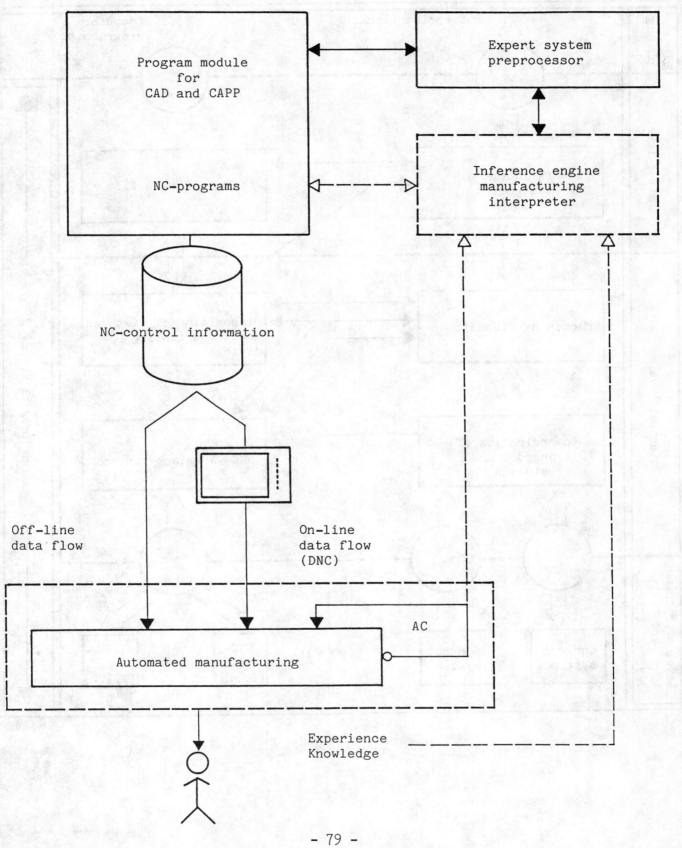

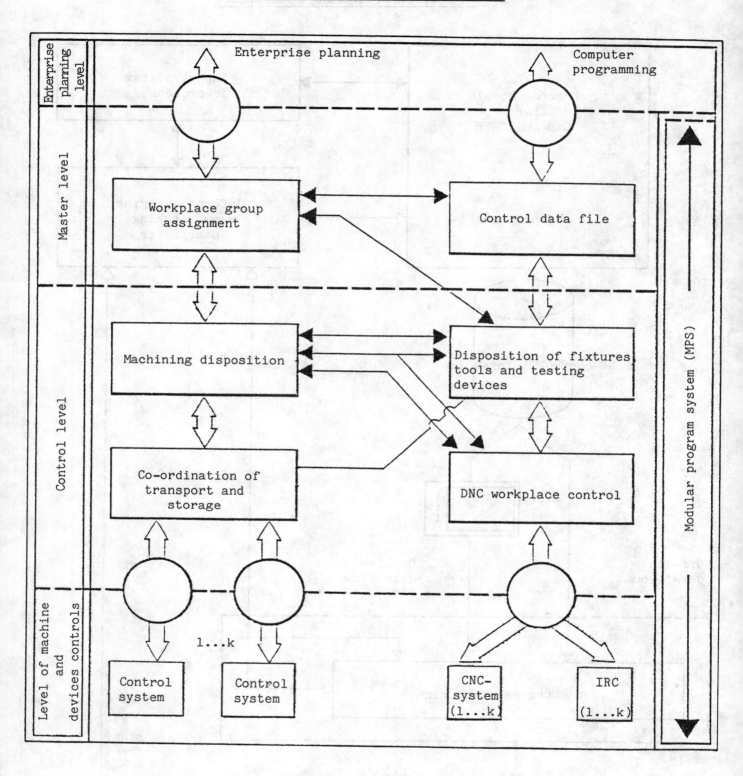

The linking of the control and scheduling systems is an essential element in the development of expert systems to handle process parameters. The feedback of parameters permits an advanced qualification of the expert system. Furthermore, the results of long-term scheduling can be assessed in comparison with significant process parameters. The process information stored within expert systems can be used for modelling the computer-aided generation of control and manipulated variables with a view to increasing process stability.

A cell computer with additional functions planned for use in manufacturing or duplex cells can be considered as an appropriate control structure to link manufacturing and monitoring equipment with intelligent scheduling systems. The structure of the cell computer consists of the following two hierarchic levels:

- Sequencing and co-ordination level - for control of the technical sequences, the manufacturing equipment and the "teaching" and presetting of monitoring devices; and

- Organization and communication level - for

    - Registration of manufacturing status information,

    - Establishment of parameter correction calculated on the basis of drifts and trends of parameters measured,

    - Co-ordination and selection of process parameters for use in the scheduling system.

The organizational and communication requirements should put particular emphasis on the design of the control structure.

DNC-communication can be used for the control of:

- NC-programs;
- Programs for handling tools and parts;
- Tool files;
- Registration of the status and progress made in manufacturing; and
- Programs and reference parameters used by monitoring devices.

By using loadable program modules, it is possible to deliver

- Specific routines for diagnostics;
- Programming and editing of routines; and
- User-specific statistic functions.

II.2.   Examples of installed systems for industrial automation

A large number of software packages for industrial automation applications are currently available in all industrialized countries.   As various examples are described in chapter II.1 above and in the annexes to the present study, only a few software packages for industrial automation in use in several countries of the ECE region and Japan have been selected for illustration in this subchapter.   They include software for CAD/CAM and industrial robots.

II.2.1.   Examples of installed systems in the United States, western Europe and Japan

II.2.1.1.   Advanced software for CAD (geometric modelling) [23]

One of the current trends in CAD/CAM systems development is an integration of subsystems with different structures and various design and technological functions around a product definition data base.   The main part of such a data base is the information which characterizes form, structure, and position of an object in space.   The process of creating and changing this information is called geometric modelling.   This process is realized by a software-hardware facility usually called a geometric modeler.

Geometric modelling accomplishes two important things:

(a)   It synthesizes the analytical definition of the geometric object with the predetermined properties in the computer;   and

(b)   It uses all numerical methods in mathematical experiments to determine and study the various properties of the designed object in the computer before the object is manufactured.

Most of the leading companies selling hardware and software for CAD provide systems with various levels of integration.   The basic system is a data base containing a three-dimensional object geometry model and information about other characteristics of the object.   This basis may be termed a "project model".   All subprocesses of integrated CAD/CAM, the analysis by the finite element method, optimization, documenting, program preparation for NC-tools, manufacturing, assembling, and robot programming, have access to the project model.   The process of geometric modelling is normally initiated by the user at the design stage, but there are also cases when it results from application programs.   Forming the model picture for the graphic device and inputting from it are performed by special application programs, basic graphic systems such as Core System and GKS.   The project model storage and data exchange between different systems can be realized in IGES standard.

Used analogue methods in CAD/CAM software systems development have the advantages of automatic problem solving, minimum requirements for computer power and easy programming.   Their disadvantages are a restricted subject field, low flexibility, program dependence on subject type and the need for preparation work in developing the generalized drawing.   Systems using these methods are SPLINK,

FAP-KF, BUILD and COMVAR.   The preferred systems are those which use the variant and search modelling methods.   The essence of these methods is that they initially give some properties and restrictions (geometric, physical, mechanical, technological, etc.) which are typical of the object considered.   They also define the class of mathematical models, describing designed objects as a system function of many variables depending on unknown parameters.

The search for these parameters is carried out by mathematical programming methods.   The advantages of these methods are the possibility of obtaining new, non-trivial solutions, flexibility, a quick change of subject field and the possibility of multicriteria problem solving.

The description of the objects is mostly realized by a three-dimensional representation.   Three-dimensional objects modelled in CAD can be subdivided into the following four classes:

(i)   Wireframes;

(ii)   Polyhedral surfaces;

(iii)   Sculptured surfaces; and

(iv)   Solid modelling.

Following is a brief review of a selection of advanced CAD systems.

The CADD system (McDonnell-Douglas, United States) can be used for the investigation of part configuration, the representation of parts, sheet elements and their modifications, assembly drawings and pipelines, and the modelling of unit translations.

The EUCLID system (France) is of interest from the point of view of language resources.   Objects of the system are described by a special language and approximated as polyhedrons.   Cylindrical, conical, and other surfaces can be described.   The EUCLID system allows definition of dimensions of complex parts on the basis of assembly drawings and representation of new parts by sectioning the existing ones.   The system is realized in FORTRAN.   It offers the possibility of analysing part repositioning and the conversational development of technology for NC-tools.   With EUCLID it is possible to represent parts whose forms are approximated as polyhedrons with plane faces.

The input language provides specification of solid elements (tetrahedrons, prisms, pyramids) as well as of two-dimensional elements (points and planes).   Two ways of specifying object form are provided for in the system.   In the first, the space position of some surfaces is specified, the boundaries of the surfaces joining each other.   In the second, the defined surface is used as the main element of the specified profile body.   The operations of set-theoretical joining and addition and subtraction of some solids are possible, giving representations of complex drawings.   The system reproduces representations obtained from various aspects.

The GEOLAN system (Messerschmitt, Federal Repubic of Germany) uses an input language analogous to APT. The language commands are interpreted. The system allows definition of any geometric objects (points, straight lines, surfaces, sections of bodies, lines of body intersections, etc.) and their representation. Parts are described by parametric curves and surfaces. With GEOLAN, part and assembly drawings are obtained automatically. The specified geometric contours are loaded into the memory and can be fully or partially reused.

The GEOMAP system (Japan) is intended for three-dimensional graphic operations. Part forms are specified by simple solid elements which have contact through certain surfaces or can intersect with each other. The system allows representation of parts and units in various positions in space and geometric modelling of mechanism kinematics.

The CLIDE system (United States) is also intended for three-dimensional graphic operations. The input language consists of commands analogous to ALGOL statements and is used to specify not only commands but also geometric definitions and programs. The system is written in BLISS. It is also used for data base management where geometric descriptions of parts are stored. By geometric definition in the input language, geometric objects can be specified. The introduction of new geometric elements with arbitrary topology or structure is possible. By the addition or subtraction of simple geometric elements, the form of complex parts is described. The language has a command for checking the collision of construction elements. Parts are represented in perspective projection or in orthogonal co-ordinates, hidden lines being marked as needed.

The CADAM (Computer-Graphics Augmented Design and Manufacturing) system (CADAM, Inc., United States) provides interactive graphics for CAD/CAM functions [97]. CADAM comprises a high-performance, high-function design/drafting package, together with a number of aids to design analysis. In addition, it provides a capability for NC part programming. The CADAM system uses a central design data base for storing and retrieving drawings. This design data base enables users in both design and manufacturng to share geometric and alphanumeric data.

The CATIA (Computer-Graphics Aided Three-Dimensional Interactive Application) system (Dessault Systemes, France) is a highly interactive, high-function 3-D geometry system for advanced CAD/CAM applications [97]. The system allows for the direct construction and manipulation of three-dimensional objects. Objects can be defined as a solid, surface, or wire frame model. The CATIA system can accept geometry from other systems or the user's own programs by means of its geometry interface. A 3-D model can be viewed from any angle automatically, with several views displayed simultaneously. The kinematics function allows motion studies and interference checking.

The systems here described have many common characteristics. All are designed for three-dimensional graphic operations. For part description they use languages, tables, or interactive modes of operation using a graphic display. Simple geometric elements are combined through contact surfaces or by inter-sectioning. Hidden lines can be specified automatically or in a conversational mode for visual representation.

The distinctions between these systems lie mainly in their differing design descriptions of part geometry and in their use of different geometric elements and ways of interrelating.

Trends towards artificial intelligence in geometric modelling systems

In practice, this means that the system itself should be able to perform new and more complicated design operations without requiring instructions from a designer on how to do so. The means of realizing such system properties are banks of geometric knowledge and a methodology of semantic models. The application of the artificial intelligence technique is possible in several directions: first, development of languages which are as close as possible to natural ones; secondly, development of expert systems that allow planning and evaluation of the solution process on the basis of a given general problem description, and that choose automatically the optimum way of achieving the solution.

The geometric modelling system should be considered a complex one, including a number of its functional subsystems with their specific informational resources:

(i)   Object formation;

(ii)   Mass properties and geometric analysis of objects;

(iii)   Modification and editing of forms and structures of geometric objects;

(iv)   Location of geometric objects on the surface and in the space;

(v)   Modelling kinematics of geometric constructions; and

(vi)   Optimization of formulation and solution of geometric problems.

All informational system resources should be accessible to a designer directly from the work-place at any time and in the required quantities. In other words, there should be easy and natural access to the system from remote graphic terminals and terminal stations through communication channels. The requirements and properties enumerated here do not cover all the problems. Nevertheless, they give an idea of what features a prospective geometric modelling system as an important component of CAD should possess.

II.2.1.2. Advanced software for CAP [23]

The following are examples of advanced software for CAP.

The advanced manufacturing, accounting, and production system (AMAPS) is an on-line, functionally complete and fully integrated manufacturing and control system. AMAPS provides information that enables management to plan and control manufacturing operations with precision and thus make the most productive use of a company's manufacturing resources. A special option enables manufacturers to incorporate multiple plants into a single data base. AMAPS is a flexible system that utilizes control file parameters to enable its users to establish management policies and procedures unique to their particular companies and industries. The system is innovative and easy to use and includes such user-friendly features as on-line MRP-type analysis and optional real-time data base updating directly from inquiry screens.

The automated time standards (ATS) module was developed by the McDonnell-Douglas Automation Company, under the sponsorship of the CAM-I Process Planning Program. ATS provides a means for proving the feasibility of automating the time-estimating process for manufacturing operations. It is a prototype software applications package which permits user-defined estimating sequences to be automatically completed with user-written calculation routines. ATS can be used by itself, or in conjunction with the CAPP system. When used with CAPP, ATS permits information in a CAPP-stored process plan to be used as input to complete an estimating plan.

CAPP is a prototype software development that provides a data-management framework designed to assist the functions of process planning in the manufacture of discrete parts. The system enables a process planner automatically to access standard process plan specification data in an interactive and dynamic manner. These standard process plans are directly related to the specific fabrication processes required to manufacture a family of parts. Once the standard process plan is retrieved, the planner can refine the plan to produce a similar part. The refined plan data can then be stored and become part of the computer data base for subsequent recall as required for production or update. CAPP 2.1A is directly implementable on an IBM 360/370 system under operating system OS/MVT, Release 21.7 with Time Sharing Option. CAPP 2.1A consists of a nine-track magnetic tape with supporting hard-copy documentation. The modifications are in hard-copy form only, with the exception of the IBM 3270 modification, which is available on nine-track magnetic tape with hard-copy documentation. A minicomputer version of CAPP is also available. CAPP was developed by CAM-I, Inc. The software has a module for the process-planning function.

CUTPLAN is a computerized process-planning system which aids process planners and manufacturing engineers in developing the manufacturing information required to produce a given part. CUTPLAN has both variant and generative capabilities for the development of both routing sheet information and detailed operation information, such as cutting-tool selection, feed/speed selection, and cut selection. Basic sets of machining data can be supplied by Metcut, or company-specific data may be incorporated in the system. CUTPLAN also has graphics capabilities for the development of manufacturing drawings such as set-up, machining, and inspection instructions. Additional stand-alone modules for machinability data retrieval (CUTDATA), tool and cut selection (CUTTECH), and cost estimating (CUTQUOTE) are currently under development. CUTPLAN was developed by Metcut Research Associates, Inc., Cincinnati, OH (United States). The software has a module for the process planning function.

DAL is an interactive computer programming language that is an integral part of the Calma DDM (design, drafting, and manufacturing) CAD/CAM system. DAL includes all of the DDM graphic commands for geometric construction, plus standard mathematical functions, display controls, labelling and dimensioning, and branching and conditional execution. DAL has an interface to FORTRAN. DAL was developed by Calma, a subdivision of General Electric Co. The software has modules for these functions: interactive design, numerical control programming, automated design drafting, digitizing and scanning capabilities, engineering analysis, process planning, material requirements planning, production scheduling and control, inspection and quality control, and inventory control.

Calma's three-dimensional mechanical design DDM system is a turnkey, stand-alone interactive graphic system.   This versatile general-purpose manufacturing tool has a number of highly automated applications, such as automated design and drafting, finite element modelling, flat pattern development, family of parts, parts nesting, flame cutting and numerical control.   In performing the numerical control functions, the system operates with the same three-dimensional model data that were used to create the original part design geometry, to automatically or interactively generate tool paths.   The system will handle such major numerical control areas as 2 1/2- to 5-axis surface contouring, turning operations and profiling and pocketing to surfaces.   The tool paths can be displayed on the graphics screen for visual verification and editing and for checks on dynamic and fixture clearances.   When the tool-path data has been verified, it can be output as a CL file, APT source file, COMPACT II, or in any convenient form designated by the user.   DDM was developed by Calma, a subdivision of General Electric Co.   The software has modules for these functions: interactive design, numerical control programming, automated design drafting, digitizing and scanning capabilities, modelling and finite element analysis, process planning, material-requirements planning, production scheduling and control.

II.2.1.3. Advanced software for industrial robots

Low-level programming languages

Programming languages for IR in the form of classic assemblers or interpreters have all the advantages of low-level languages, permitting optimum coding with a minimum of redundances.   Their disadvantage is that they tend to be orientated to one type of robot control unit and are thus incompatible with the software of other types of robots.   Programming languages of this class are used at: ANOMATIC from Anorad Corporation, SUPERSIGMA from Olivetti and IRb/60 from ASEA.

Producer	Anorad Corp.	Olivetti	ASEA
Type of robot	ANOMATIC	SUPERSIGMA	IRb/60
Language	ANORAD	SIGLA	Untitled
Processor	M 6800 Motorola	Microcomputer Olivetti	I 8008 Intel
Basis of language	NC progr.	Assembler	Assembler
Type of language	Interpreter	Interpreter	Dialogue
Hand control	No	Yes	Yes
Number of instructions	20	35	16
High level commands	Yes	Yes	No
Interaction with environment	Yes, I/0	Yes, sensors; I/0	Yes, sensors; I/0
Stepping	Yes	Indirectly	Yes
Editing	No	Yes	Yes
Calling of subroutines	Yes	Yes	No
Compatibility with other robots	No	Yes	No

The information provided in the above table shows the basic parameters of programming languages of this class. It may be seen that the capabilities of these languages are more or less identical.

High-level programming languages

(a) Structured programming level

Structured high-level programming languages control the complex motions and functions of the robot. In most cases they permit the application of transformation, the description of the robot environment and the definition of states variables. Parallel programming and a high degree of environment interaction is provided. The following are examples of programming languages which may be included in this class:

AL   – Developed in the Artificial Intelligence Laboratory, Stanford, CT (United States). The language is derived from Algol and can define data types (scalars, vectors, rotations and positions). Operations over these types of data are defined and motion transformation with reference to a hand-held object is possible;

MCL   – Developed by McDonnell-Douglas on the basis of ATP, well known for the programming of NC-machines; and

MAPLE – Developed by IBM. The language is based on the high-level language PL/1, supplied with special features for robot control.

Robot manufacturing firm	Stanford	McDonnell-Douglas	IBM
Language	AL	MCL	MAPLE
Processor	PDP 11/45	Microcomputer	IBM Syst.7
Language based on	Algol	APT	PL/1
Type of language	Compiler and Interpreter	Compiler	Interpreter
Structure	Begin/End While/Do/If/ Then/Else	When/Else End of While/End of	Go to While/Do If/ Then/Else
Calling of subroutines	Procedures Functions Macros	Macros	Procedures
Transformation of co-ordinates	Yes	Yes	Yes
Contact sensors	Yes	Yes	Yes
Visual system	No	Yes	No

These languages replace programming of the PTP type by operation descriptions with more complex motion definition. The possibility of describing the robot's working environment is also provided, as well as the modelling of conflictless operations. A considerable advantage is the possibility they provide of modelling outside the robot's control system, i.e. "off-line" creation of the manipulation programs. The need for the presence of a skilled programmer, acquainted with the robot's application area, may be considered a disadvantage.

(b) Special problem-oriented languages

Special problem-oriented programming languages are designed to minimize the demands of programming and yet, at the same time, to retain all the powerful properties of high-level languages. However, these types of programming languages are still in the early stage of development. One of the first languages of this class is AUTOPASS developed by IBM. The language uses high-level commands similar to those normally used by a man during assembling work. For instance, the command PLACE object X from position 1 to 2 includes the location and identification of object X in position 1, the definition of motion from position 1 to position 2 and the placing of the object in position 2. As may be seen from the example, environment modelling, visual seeking and object identification must be implemented when using the AUTOPASS language.

A different approach to robot programming on that level is employed by ASEA for their robots equipped with a third-generation control system. It is based on operator/system dialogue and a special operator panel has been designed for the purpose. The system can be supplemented by a camera and a screen for object identification. Again, the programming is automated in the sense that the system asks the operator for specification of the object points if it is not able to differentiate them itself.

Special problem-oriented languages represent the adanced trend of robot programming. But their high level requires environment-description modelling, artificial-intelligence methods for decision-making and an interactive debugging system. It is evident that the development of programming languages of this class is demanding in respect of time and human and financial resources. However, in spite of these disadvantages, the development of industrial robot applications will not dispense with the programming systems of that class.

Methods of off-line programming for IR

The current highest level of IR-Programming is called off-line programming, where robot programming may be seen as a task of technological planners using techniques available for the description of manufacturing systems [51].

The Institute of Production Systems and Design Technology (IPK-Berlin-West) is currently engaged in the development of a prototype of an off-line robot programming system. The aim of the work is to transfer the programming activities to the technological planner and to ensure that the programmes thus developed - with minimal changes - can run in the production process. Figure 31 describes the conceptual structure of such a system. The central modules are the IR-program generation with which the user can develop application programs and the

Figure 31. Structure of an off-line robot programming system [87]

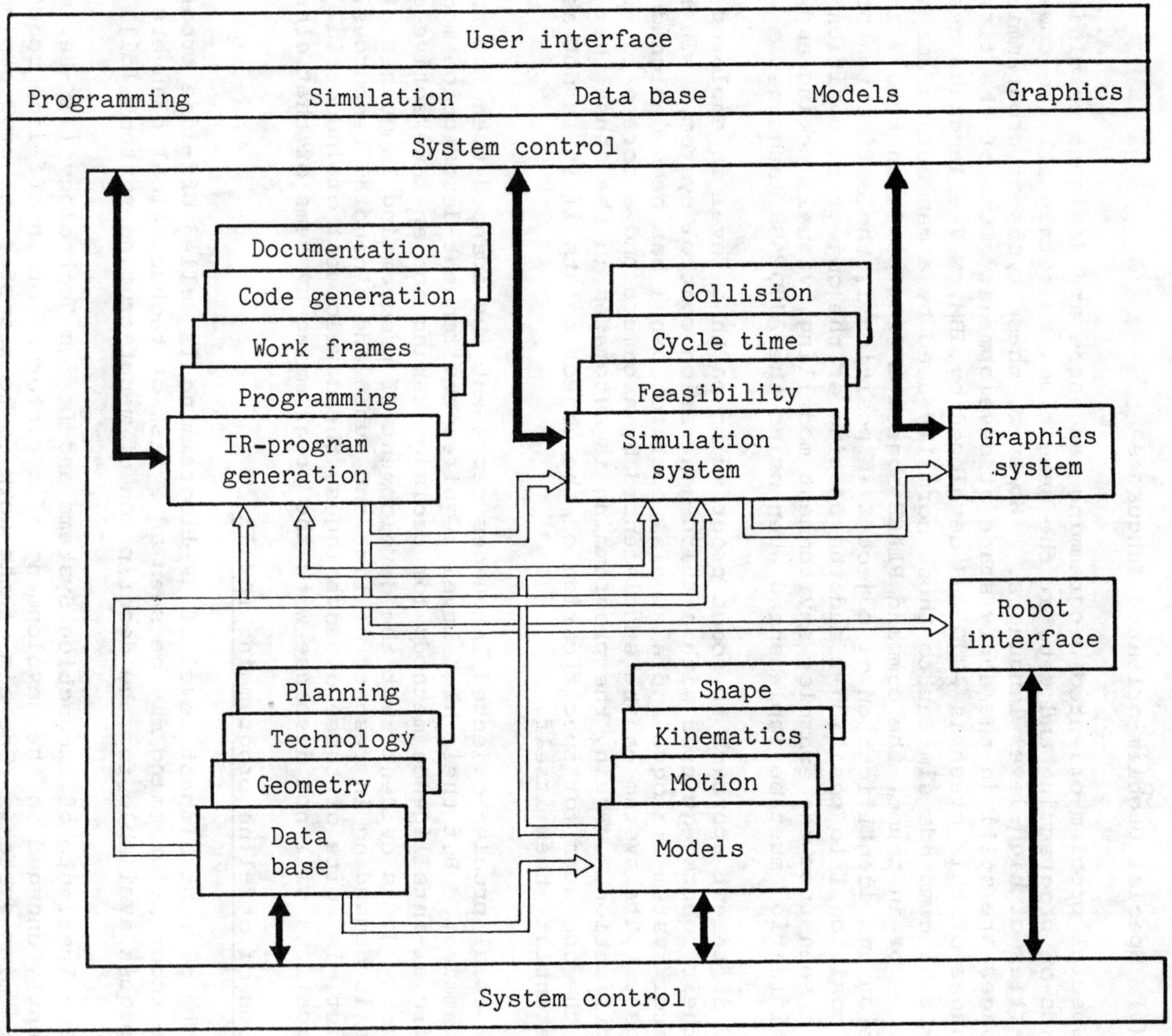

simulation system which can be used to test off-line generated programs and to generate the time span required for the execution of given tasks. Testing implies the checking of defined paths concerning position, orientation and speed of end-effector statements and interactions with peripherals. It is also necessary to check the availability of the predefined robots, tools, workpieces etc., and the physical environment of the work place.

All components used must be simulated with their functions. The two function units - IR-program generation; and simulation system - have access to data bases and simulation models. The user interface makes functions available to the user who can communicate with the system using alphanumeric and graphic means. Via an interface, the off-line generated and tested programs are made available for application in the real robot.

Graphic simulation is a tool which can be applied effectively for off-line programming tasks. It is expected to be particularly effective in preventing collision. But the need for numeric simulation remains and it is necessary to image the control characteristics in the data base of the system as well as the motion model, the kinematic model and the shape model of the robot, the tools, the conveyors, the workpieces, the fixtures and the environment. The user interface and the interfaces between the programming system and the industrial robot must be recognized for the layout of the system architecture. The information concerning models, cells, tests, programs and scenario must be available in the data base for systems control. The efficiency of generated robot programs is highly dependent upon an exact simulation of the environment. It is therefore necessary to provide a mathematical description and a computer internal representation of the geometry and kinematics of the robot and its environment.

II.2.2.   Examples of CAD/CAM-solutions in eastern Europe

In all the countries of eastern Europe, many CAD/CAM systems are under development and in practical application. The most important in this field are general frame systems. The German Democratic Republic has available the RATIBERT system, which was originally a CAPP system. Following further development, it will be well suited as a CAD/CAM frame system, supporting numerous user functions.

The following basic functions of the data-handling process are supported by modules of this system:

- Loading (establishing) of data stocks;

- Replacing, clearing, blocking, checking, selecting, compiling, restructuring and adding of data stocks, data records and data;

- Output (print-out or display screen) of data stocks or records, complete and with the use of variable group keys;

- Direct access to data arrays, extraction of data, combination and averaging of data sets, and conversion and supply of data for computations;

- Searching in data stocks and data records; and

- Rearranging and securing of data.

- 91 -

The term data stock refers to a group of data records having the same or different structures and with definable relations to objects. In a borderline case, a data stock may also consist of a data record. In the storage structure only one address is assigned to the data stock, so that it can be located by a single access command. Not only the work-plan master cards and lists of parts (consisting of records of different structures) are imaged in the data and storage structure as data stocks, but also data collections, e.g. texts, workplace data, material data and normative rules.

Data records constitute data and relationships which can be imaged and interpreted in a variety of structures. The RATIBERT system can be used at all conceivable intermediate stages of development, from a basic system for the computer-aided treatment of the work-plan master card up to a complex information processing system which integrates planning, design and manufacturing processes.

The system can be supplied for a variety of input and output modes, such as punch-card automated processing cycles and display-screen dialogue and can comprise facilities for:

(i)   Working out of the sequences of operations and the calling in of production-related source programs on the basis of

  - Dummy and type processes

  - Sequencing graphs

  - Generative process programs

  - Similarity techniques;

(ii)   Treatment of the work-plan master cards and operational instructions;

(iii)   Job-controlled and group-process-controlled generation of documents for production;

(iv)   Determination of price and cost for every new part and in case of change concerning production technology, design, and blanks;

(v)   Computer-generation of data and an automatic updating service for the data bank of the enterprise;

(vi)   Provision of data on savings in time and material for planning and accounting purposes;

(vii)   Determination of optimum lot sizes and distribution modalities;

(viii) Computerized identification of repetitive parts, creation of similar parts, and geometric (or detail-geometric) analysis of parts, including:

  - Automated design and preparation of a variety of alternatives with all economic parameters on the basis of contract specifications

  - Economic evaluation of alternative designs

- Provision (in an interactive computer mode) of production
  information via a video terminal (assortment plan, main plan of
  dates, material use, workplace, indication of finished production,
  etc.)

- General computer-aided updating service for all stored data stocks

- Execution of any scientific-technological computations;

(ix)   Processing of programs for

- Automatic programming of NC-machines

- Preliminary computation of time for conventional centre-lathe
  operation

- A large number of conventional operations (peripheral milling, flat
  grinding, internal grinding, heat treatment).

The development of CAD/CAM systems in Czechoslovakia began already in the
early 1970s.   Since 1978, the SAT system has been in regular use for the
manufacture of blades for turbines and axial compressors.   The system was
presented for the first time at the United Nations Economic Commission for Europe
Seminar on CAD Systems as an Integrated Part of Industrial Production, held at
Geneva in 1979.   The first version of the system worked in off-line mode and the
interactive on-line version has been available since 1983.   It has been useful in
the establishment of scientific calculation libraries and in the development of
finite and boundary element methods.

Since the beginning of the 1980s, several extensive CAD/CAM systems have been
developed, oriented at specific product fields.   CAD/CAM systems are applied on a
high level in the design and manufacture of electric machines (electric motors,
transformers, etc.), where traditional computer interactive alphanumeric control
techniques are used to execute the calculations.   In the implemented systems,
graphics techniques are employed largely in a passive way.

In Czechoslovakia, work is at present under way on the development of CAD/CAM
systems employing interactive graphics techniques.   CAE systems are currently in
the stage of preparation.

The development of large-scale data-base-oriented industrial control systems
is complicated, time-consuming and costly.   Using the traditional development
approach, designers must take decisions already in the design phase regarding the
functional data and the structure of the operational system.   The quality of
these decisions becomes apparent only at the time of system implementation, one to
two years later.   On the basis of first implementation experience, the system
usually has to be modified or reworked in order to meet user and operational
requirements.

By using the prototype approach, it is possible to avoid the cost of
redesigning.   In this case, a prototype of the target system is constructed,
based on systems design specifications.   By means of SAD (software for automated
design) tools designed in Czechoslovakia, it is possible to compile such a
prototype system in one to two weeks, so that target system feasibility simulation
and user acceptance can take place in the early development phase.   The above
prototype approach has been successfully implemented in three production control
projects in Czechoslovak engineering industries.

Further examples of software for industrial automation available in eastern
Europe are presented in annexes II to VIII.

II.3. Current trends in software for industrial automation

At present, there are two trends in international standardization for industrial automation which are of particular importance. These are the work being undertaken towards standardization of a neutral format for the exchange of product models in CAD/CAM systems (see chapter II.1.6) and towards communication standards for CIM structures (see chapter II.1.7), by means of which the total integration of the informational network should be promoted.

II.3.1. Product models

CAD/CAM data exchange is concerned with the need for industry to transfer information between dissimilar CAD/CAM systems. However, CAD/CAM data exchange is fraught with difficulties. For example, there exist:

- Over 200 suppliers of CAD/CAM systems - few of which are compatible in terms of their ability to communicate, or in terms of the industries and exchange standards they support;

- Many different data exchange standards - some with national or industry-specific affiliations;

- Pressure from many of the largest multi-national companies, whose demand for automated data exchange is ever more insistent; and

- Shortage of unbiased, accurate and up-to-date information on the subject.

To overcome the difficulties, efforts have been made in several countries, on the one hand, to create a neutral format for the exchange of product models, and, on the other hand, to disseminate correct and timely information on the results achieved. 1/

International development in the field of product models has taken place in the following stages:

1/ In the United Kingdom, for example, a CAD/CAM Data Exchange Technical Centre has been created in Leeds with a view to publication of authoritative works on data exchange (and, in particular, the IGES standard), to provision of software, test reports, advice and support to ease the exchange of information between differing CAD/CAM systems, to education of companies' staff, suppliers and customers in the complexities of data exchange, to influencing the process of creating standards nationally and internationally, etc.

September 1979   - IGES development within the framework of the ICAM project

1980   - Publication of IGES version No. 1

1981   - IGES-based ANSI standard accepted (co-operation amongst 60 United States firms)

July 1982   - IGES version No. 2

May 1985   - IGES version No. 3.

This development was confronted with the following general problems:

- The specification is difficult to understand and somtimes ambiquous;

- Vendor support from IGES Committee is insufficient, for example, in the areas of implementation quidelines, explanation of the specification, development of testing and validitation procedures;

- The IGES files are large compared with CAD data files; and

- Vendors' processors implement ill defined subsets.

Therefore, the French aeronautical industry together with MBB (Federal Republic of Germany) and the aeronautical industry of the United Kingdom have pursued their own standard-development, which is of international importance.

January 1983   - SET (System for Exchange and Transfer)

March 1984   - SET version 1.1 (French)

September 1984   - SET version 1.1 (English) with interfaces for

· CADAM
· Computervision CADD S3
· CDM 300

This development is being promoted in western Europe.   The data exchange with SET for drawing information was carried out for the Airbus 320 on a large scale.

Some of the technical projects that have contributed to the history of the IGES project and possible future developments in data exchange specification standards are shown in figure 32.   It is to be hoped that the many data exchange

Figure 32. Development of data exchange specifications [88]

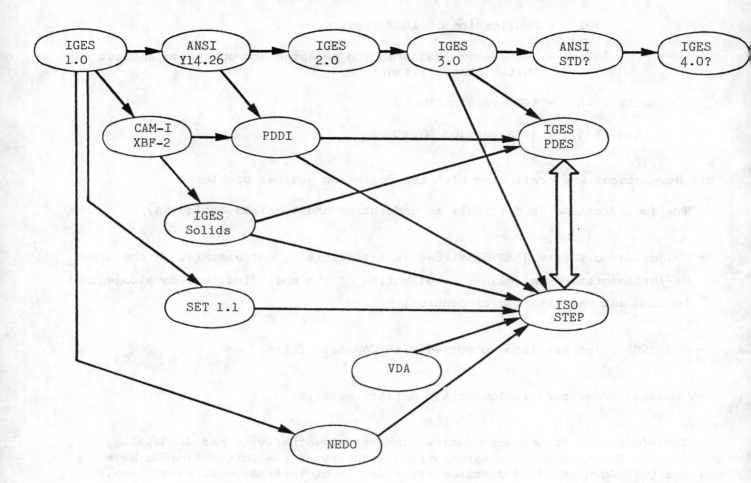

IGES   = Initial graphic exchange specification

ANSI   = American National Standards Institute (United States)

CAM-I   = Computer-aided manufacturing - International (International Organization)

NEDO   = National Economic Development Office (United Kingdom)

PDDI   = Product data definition interface

PDES   = Product definition exchange standard

STEP   = Standard for exchange product definition data

SET   = System for exchange and transfer

VDA   = Association of car manufacturers of the Federal Republic of Germany

formats under development will eventually lead to a world-wide ISO standard
(STEP).   By the product definition exchange standard (PDES/STEP), methodical
assistance for practical use is made available.   The first version of STEP gained
preliminary international agreement in March 1985.   The aim is the creation of a
first version for industrial application by 1990.   The two temporary solutions
currently being used will continue to be applied until that date.

II.3.2.   The MAP and TOP communication standards

The need for communication standards and their basic principles have already
been explained in chapter II.1.7 above.   The following paragraphs therefore limit
themselves to the essential steps of the development of those standards.

The MAP (Manufacturing Automation Protocol) standard design, proposed by
General Motors and accepted by leading United States firms and by some west
European firms such as Siemens, is developing into a de facto standard for all
manufacturing industries.

MAP is a broadband, token-bus communications protocol based on the Open System
Interconnection (OSI) model established by the International Organization for
Standardization (ISO).   The ISO model identifies everything that must take place
before intelligent devices, such as programmable-logic controllers, CNC systems
and computers can communicate with each other.   Currently, a pilot project is
testing the communicating devices under actual conditions.

MAP standards have initially been applied to painting robots.   Their
implementation then spread throughout the pilot installation of General Motors at
their Detroit-Hamtranck facility, covering all their 2000 programmable devices
installed there.   The development and application schedule of MAP is presented
below:

1978   ISO published the Open Systems Interconnection (OSI) model.
                       Since then, it has been almost universally accepted as a
                       pattern for local area network (LAN) development in both
                       factory and office.

1980   The Institute of Electrical and Electronic Engineers of the
                       United States (IEEE) created the Project 802 committee to
                       start work on LAN standards.

April 1980   General Motors established an internal task force with
                       representatives from seven divisions for the aim of
                       developing a communications standard to permit communication
                       amongst programmable devices from various manufacturers.
                       This led to the development of the Manufacturing Automation
                       Protocol (MAP) based on IEEE standards and the OSI model.

June 1982   General Motors formally adopted MAP as a communications
                       standard for all its plants and requested that equipment
                       suppliers should follow the MAP standards for interconnection.

August 1985	General Motors implemented a preproduction pilot MAP installation at their Detroit-Hamtranck plant.
September 1985	The MAP Users Group became an official Technical Group of the Society for Manufacturing Engineers in the United States (SME).
November 1985	General Motors and Boeing co-sponsored a major demonstration at the Autofact show in Detroit, with MAP interconnected with TOP (Technical & Office Protocol, developed by Boeing). The resulting MAP/TOP network demonstrated the feasibility of a multivendor, computer-integrated manufacturing facility (figure 26). More than 21 companies participated.
Spring 1986	General Motors to implement MAP in pilot projects at their Saginaw Steering Gear Division and at B-O-C Hamtranck. Operational systems will be installed at three Truck and Bus plants.
1989	General Motors will start production of their Saturn model automobiles, which will be made with MAP as the exclusive communications protocol.

A specific goal in the development and application of MAP is to adhere to the networking protocol structure defined by ISO. This model for Open Systems Interconnection (OSI) is supported by most national standards organizations.

The highest layer of MAP is occupied by the application programs and the lowest consists of the physical media over which the data is transmitted. An overall view of the model with a description of the assigned layer functions may be found in figure 33. The model allows the use of multiple sub-layers to implement a given layer or that of zero layers where given functions are not required.

The network layer is concerned with the routing of messages between the two (or more) communicating applications in the presence of intermediate nodes. Protocols at this layer must be consistent throughout a given network. The definition and implementation of the network layer is crucial to the success of MAP as a non-proprietary, multi-vendor network standard.

The Data Link and Physical layers are concerned with the physical communication of messages between adjacent nodes. Protocols at these layers must be consistent within the connection of each pair of adjacent, communicating network nodes. The use of MAP standard protocols at these layers will allow for the interconnection of nodes from different vendors.

Each layer provides services to the layer above. Two types of information are passed between layers in providing these services: control and data. The control information is the basis for all the services which are required to process the message. As each layer provides its part of those services, the remaining control information is passed to the next lower layer. This process continues until no control information remains.

Figure 33. Open systems interconnection (OSI)
reference model [89]

Figure 33. Open systems interconnection (OSI)
reference model [89]

Layers	Functions	Layers
User programme	Applications programs (not part of the OSI model)	Server machine
Layer 7 application	Provides all services directly comprehensible to applications programs	Layer 7 application
Layer 6 presentation	Restructures data to/from standardized format used within the network	Layer 6 presentation
Layer 5 session	Name/address translation, access security, and synchronized management data	Layer 5 session
Layer 4 transport	Provides transport, reliable data transfer from end node to end node	Layer 4 transport
Layer 3 network	Performs message routing for data transfer between non-adjacent nodes	Layer 3 network
Layer 2 data link	Improves error rate for messages moved between adjacent nodes	Layer 2 data link
Layer 1 physical	Encodes and physically transfers messages between adjacent nodes	Layer 1 physical

Physical link

The data which are passed down to a layer are generally transported transparently. (An exception is the presentation layer whose task it is to reformat that data.) Each layer prefaces the data with a control block prior to requesting the services of the next lower layer. This control block is interpreted by the corresponding layer in the receiving node.

As the data is passed to successively lower layers, its size increases, as shown in figure 34. As the data is passed to successively higher layers at the receiving node, the control blocks are removed.

MAP has relatively high requirements on the hardware of programmable control units. These call for a memory capacity of about 500 KB, internal operating systems and appropriate communicating hardware.

To ensure complete integration of a manufacturing business it is also necessary to bring together the technical and office functions. For example, electronic mail, word processing, text interchange, file transfer, graphics, videotex, database management, business analysis and product data. Here the complement to MAP is called TOP (Technical and Office Protocol). Like MAP it is based upon the ISO 7-layer OSI model. The current version of TOP is version 1.0 but version 3.0 will be available in 1987, alongside MAP version 3.0. As General Motors have been leading the MAP effort, Boeing Computer Services are leading the TOP activity. TOP and MAP are compatible. Both are based upon the same set of core protocols and can freely pass data between the networks. This compatibility will continue due to the close working relationship between the TOP subcommittees and the MAP subcommittees. While MAP responds to a highly structured, controlled factory environment, TOP involves information exchange in a more complex, free from environment (such as an office). TOP deals with more fragmented computing systems. Users and vendors had to narrow down and focus the options. The TOP Users Group was formed December 1985, at their first meeting in Seattle, Washington. Over 220 individuals, representing United States and foreign companies, attended the seminars and participated in working sessions addressing specific areas of interest to TOP. Based upon data gathered at this meeting, the structure and subject areas of the TOP Technical Subcommittees were defined. Within Europe two User Groups have been established: EMUG (European MAP User Group) and OSITOP (The Matching European TOP User Group).

In the United Kingdom the Department of Trade and Industry set up a practical demonstration of computer integrated manufacturing (CIM), using MAP at the NEC Birmingham in December 1986. This 5 day event called CIMAP was the worlds largest ever working demonstration of CIM using MAP and TOP. CIMAP was an awareness and education event whose aim was to show industry precisely what could be achieved by the new communications technology now. The demonstration comprised 15 cells all connected by the same communications network.

Vendors of MAP/TOP products need to have their products conformance tested. In the United Kingdom, the Networking Centre at Hemel Hempstead has established the first commercial conformance testing facility outside the United States; until recently the only recognized testing centre was at ITI, Ann Arbor. The aim initially is for the Networking Centre to provide a service identical to that from ITI for MAP version 2.1, and taking advantage of the common requirements, for TOP version 1.0.

Figure 34. Nesting of layer protocols [89]

Beginning of message on media

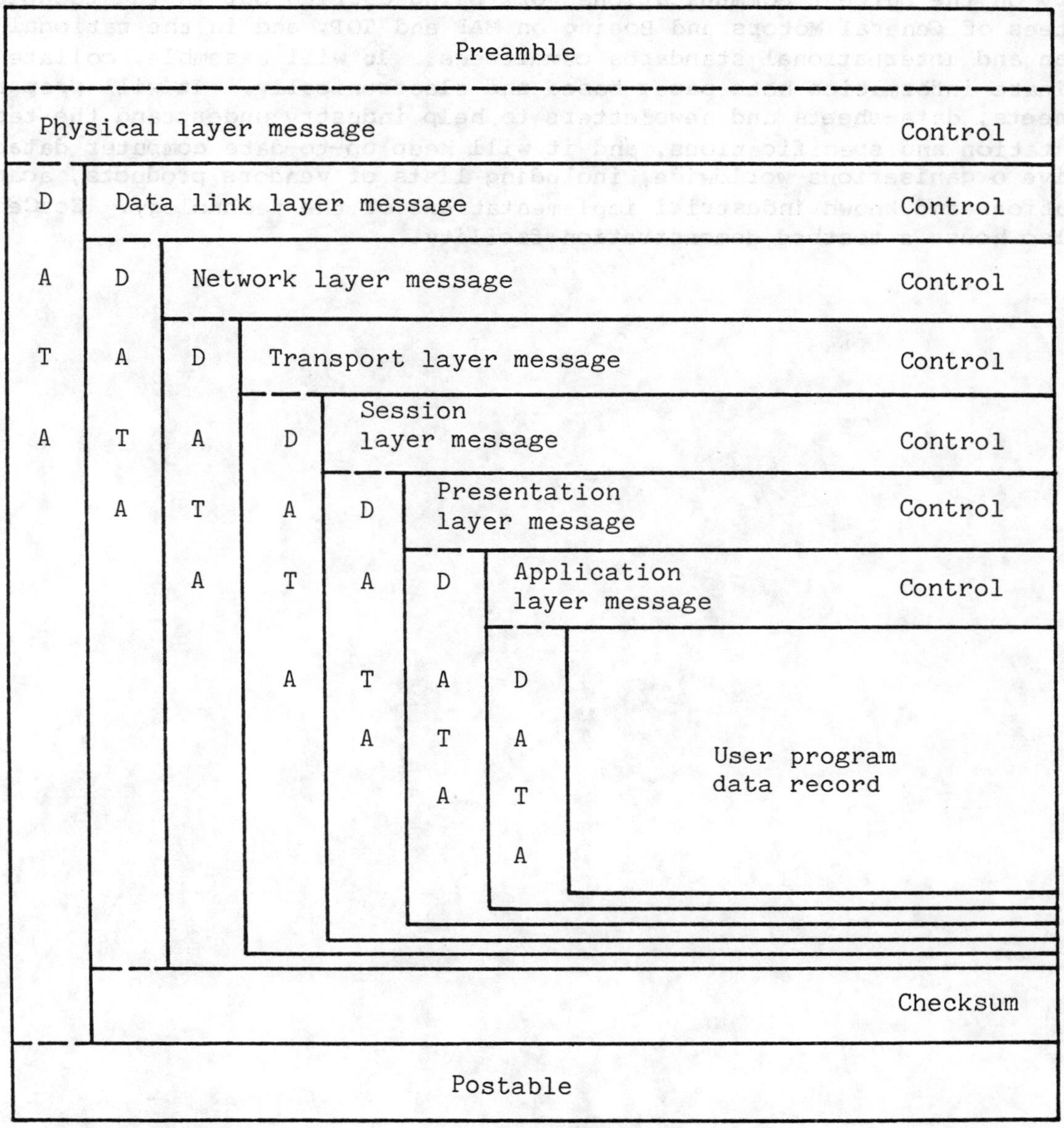

Preamble	

| D | Physical layer message | Control |

| D | Data link layer message | Control |

| A | D | Network layer message | Control |

| T | A | D | Transport layer message | Control |

| A | T | A | D | Session layer message | Control |

| A | T | A | D | Presentation layer message | Control |

| A | T | A | D | Application layer message | Control |

| A | T | A | D | User program data record |

Checksum

Postable

End of message on media

Having increased industry's awareness of MAP and TOP through CIMAP it was clear that a longer-term focus was needed to provide independent advice and up-to-date information. Therefore an organisation called ComCentre has been set up in the United Kingdom. ComCentre will advise manufacturing and process industry on the current communications work being carried out in the technical Committees of General Motors and Boeing on MAP and TOP, and in the national European and International standards committees. It will assemble, collate and disseminate information both paper-based and electronically. It will prepare fact-sheets, data-sheets and newsletters to help industry understand the technical documentation and specifications, and it will keep up-to-date computer databases of active organisations worldwide, including lists of vendors products, academic institutions and known industrial implementations of the technology. ComCentre will also house a testbed demonstration facility.

III. THE INTRODUCTION OF INDUSTRIAL AUTOMATION WITH EMPHASIS ON SOFTWARE: TECHNO-ECONOMIC, SOCIAL AND RELATED ASPECTS

III.1. Planning and organization of the introduction process

A technical prerequisite for successful implementation of integrated automation is the automation of the main stages of manufacturing (see chapter II). Furthermore, it is necessary to undertake a critical review of the product development process and the process technology applied. The CAD/CAM system to be introduced must incorporate the latest technological know-how. This is even more important than the installation of powerful hardware. Automation should start with those parts of the process where high flexibility is required. Another prerequisite for the successful implementation of integrated automation is the availability of compatible software. Furthermore, integrated automation can only be achieved after an in-depth analysis of all the existing data and of their functional relations, with a view to streamlining the network of information and reducing to a minimum the stock of necessary data to be stored in the central data bank.

Software solutions require prior complex process analyses. The omission of this important step will result in existing organizational deficiencies being incorporated into the software solution to be elaborated. The logico-mathematical model of the industrial process should both incorporate empirical knowledge and be built on the basis of technological rules/laws governing the process concerned. Thus, modelling of the process to be automated should be used to improve the process and the resulting software should incorporate those improvements (see also chapter III.4). By taking full advantage of the opportunities of modelling, risky and expensive trials on the shop floor can be minimized.

The development of efficient and accurate models and algorithms is an investment in the future of the software and thus of the total automation. The total time taken to develop a completely integrated automation system is usually about five years - something which is generally underestimated.

The software must be capable of operating under both regular manufacturing conditions and adapting to emergencies. The success of integrated automation depends on the clearly defined hierarchical classification of responsibilities and tasks in the management process. The tasks should not be adjusted to the existing management structure, but vice versa, with the management structure being adapted to the tasks. With the advent of integrated automation, traditional departmental thinking must be abandoned.

As part of the introductory process, particular consideration must be given to the qualifications of the staff and operators. Previously, the quality of the software was frequently more influenced by the specifications as set out by management than by the requirements of those responsible for developing the software, i.e. systems analysts and programmers. However, the development of integrated automation systems calls for the formation of interdisciplinary teams. There should be close co-operation between industry and university in both the elaboration of the software and the development of skilled staff. Furthermore, university graduates are often the pioneers of progress in industry.

The training of staff is of vital importance at all levels, from the skilled worker up to the enterprise manager. Provision for training should already be taken into account during the preparation of industrial automation projects. It has been found that, particularly as concerns management, the availability of PCs can be very useful in overcoming reluctance to work with the use of computers. The focal point in the provision of training is the engineering staff. In this context, increasing emphasis should be put on relevant training at the work-place (see chapter III.4). The introduction of software for automation should be carried out on the basis of an in-depth analysis of both the human and physical conditions associated with the process to be automated.

Software for industrial automation should be written in a high-level language in order to facilitate its application by staff and its future maintenance.

All considerations involving the introductory process must be set out clearly and in detail. They should cover, for example:

- An agreed plan for the gradual introduction of the automation project;

- Agreements covering assignment of responsibilities and tasks between industry and scientific institutions;

- A plan for staff training; and

- Detailed specifications for changes in work sites.

Experience has shown that the development of the software is decisive in the success of any automation project. But it is not the only aspect, because the most complicated stage in the introduction of industrial automation - e.g. a CAD/CAM application - is the process from integrated design in product development to final production. In this context the focal points are production planning, the organization of the integrated process and the training of all persons involved.

III.2. Cost-benefit considerations

Investigations of the efficiency of the CAD/CAM software are of value only in measuring the efficiency of the total solution (integrated systems solution). In determining the efficiency of CAD/CAM solutions, the following factors should be taken into account:

- The initial development and implementation costs;

- The change in the overall profitability of the factory;

- Labour saving and/or productivity; and

- Return on investment. 1/

These factors can be given different weight, depending on the objectives to be achieved.

First objective: Increased labour productivity in design and production planning as reflected by labour saving, increased labour productivity, increase in production planning capacity.

Second objective: Cost reduction in technical production planning, reduced time taken in production planning.

Third objective: Cost reduction in manufacturing (emphasis on cutting of material costs), material saving, reduction of waste.

Fourth objective: Attainment of return-on-investment targets, increase of profits.

Fifth objective: Reduction of periods of development and preparation for manufacturing and of lead time.

Methods used in the determination of these objectives and variants thereof are described in [58]. In addition, possible structures are proposed for the application of computers, of programs and of the data base. A basic variant must be selected and compared with the other variants in accordance with the following steps:

(i)   Comparison of technical, organizational, economic and ergonomic characteristics of the basic and the alternative variants;

(ii)   Evaluation of the variants and first pre-selection on the basis of the results;

(iii)   Selection of the optimum variant with respect to the main objectives; and

(iv)   Determination of efficiency increase of the optimum variant in comparison with the basic variant.

_____

1/ Attention has to be paid, however, also to social and labour implications of the application of advanced industrial automation technology. In this regard, mention should be made of studies currently undertkaen by the International Labour Organisation: "Social and labour implications of CAD/CAM (issued in spring 1987) and "Social and labour implications of artificial intelligence" (under preparation).

The growing complexity of CAD/CAM solutions is giving rise to the need to identify parameters which can be used to help determine efficiency. However, there are many parameters which cannot be quantified as yet, for example:

- The ability of a system to provide up-to-date status information;

- The availability of complete information on the total process;

- High transparency through improved organization of the manufacturing sequences;

- Reduction of paperwork;

- Increased reliability of manufacturing; and

- The possibility of simulation and evaluation at production-planning level.

There is a tendency for expenses to show an increase in the first months following the introduction and first application of CAD/CAM. The system's efficiency may be impaired in the following ways:

- Where scientific analysis, modelling and elaboration of algorithms connected with design, production planning and manufacturing are carried out before the conceptual stage, system performance may not meet expectations;

- The underestimating of co-ordination problems at the development stage may result in insufficient time being devoted to an in-depth analysis of the priorities and objectives;

- Insufficient time may be allowed for pre-implementation testing of a system; and

- There may be a lack of experience and information concerning the implementation of flexible automation and computer-integrated manufacturing.

III.3 **The changing employment structure. Training and further training of the personnel. Education and further education**

Integrated automation results in a fundamental change in the structure of professions and activities. In the production process, there is a move from machine work to programming, and from direct production to, _inter alia_, repair, maintenance and supervision, requiring highly qualified, specialized and flexible staff.

The activity of production planning is also radically modified. Information processing techniques as new working methods cause changes in the structure of activities rather than in that of the professions themselves. New professions are also created associated with activities related to software, administration and maintenance.

Staff dealing with software may be broken down into two groups: software system developers (software engineers) and software system users. Software engineers must possess expertise in the following areas:

- 106 -

- Mathematics/natural sciences on subjects such as:

    Computer geometry and computer graphics; and
    Finite elements method etc.;

- Information technology, covering:

    Programming languages and software engineering;
    Data structures and data banks;
    Computer systems and information systems;
    Simulation techniques; and
    Artificial intelligence/knowledge processing etc.

- Information and automation techniques/microelectronics; and

- CAD/CAM fundamentals and applications.

The software user must also be qualified in many areas. Provisions should be made for training in the effective use of interactive computer techniques, graphics, text processing, data handling and other software-based computer services.

The use of interactive computer techniques is a challenge to the creativity of the engineer, whose time is no longer occupied with the formal tasks now taken over by the computer. The dialogue technique makes demands on and promotes the development of the user's substantive intellectual qualities.

Radical changes in the activities structure have of course had consequences in the area of training and further training of personnel at all levels. On a world-wide basis, it is estimated that only 25 per cent of universities provide training in the field of information technology; there is therefore a need for industry to increase considerably its investment in this area. Of total training time, at least 50 per cent should consist of practical training on the computer.

As the demand for applications programmers increases, more and more large program systems for training purposes (education and training programs) are required. However, the pedagogical and psychological problems are often underestimated and efficiency is sacrificed as a result.

The demand for systems developers is characterized by two opposing trends. Thus, although it can be observed that more and more sophisticated systems are being developed, the development effort associated with systems programming as compared with applications programming is decreasing (see chapter III.4). In addition to the development of abilities and skills in interactive computer techniques, increasing emphasis is on the teach-yourself method during the work process.

Further training over the long term may be undertaken in co-operation between industry and universities. While so-called reference centres have been successful in providing research, training and further training, these centres can only meet the demands when fundamental research is carried out in conjunction with applications-oriented research. Additional provision should be made for further training of management staff, system users and systems developers.

III.4. Software provision

When developing software, attention must first of all be paid to its fundamental characteristics. The development of software is an intellectual exercise which includes the following stages (the software life cycle):

- Systems definition;
- Functional design, specification;
- Development; and
- Operation and maintenance (see chapter I.5).

The fundamental demands made of software are that it should be modular, portable, efficient, reliable and robust. The ultimate acceptance by the user (final user) depends on the functional quality of the software, as reflected by its ability to meet the above-mentioned demands.

The structuring of the software and its modularity result in a hierarchical set-up with the following basic principle: changes in the lower level should not affect the levels above.

The software's price and value do not necessarily coincide, because the value depends on the know-how incorporated in the software.

Current problems concerning software are influenced by a number of factors. At present, software depends largely on the operating system and, consequently, considerable difficulties arise as regards its portability. In the case of in-house development, sophisticated applications software is usually elaborated on a low language level, with the consequence that such software can be maintained only by its developer. Software has therefore to be developed anew for each application. Furthermore, there are as yet no generally accepted recommendations for the interfaces (definition - see annex I) between the user and the system, which gives rise to problems with system input and output commands. In addition, difficulties may arise from the fact that some programming languages have non-compatible "dialects".

One trend in software engineering which reflects the division of labour between man and machine is the change of language generations [56] (see figure 35). Current requirements call for software offering more degrees of freedom for the final user, which may be achieved by problem-adjusted programming in the interactive computer mode. Thus, in the fourth generation language, data handling is transferred from the physical to the logical level, considerably relieving the user. It is a characteristic of fourth-generation computer language that systems can be used independently of the operating system and include both data banks and languages.

Current research focuses on integrated software systems with elements such as:

- Data banks;
- Graphics; and
- Text processing.

Figure 35. Division of labour between man and machine depending on the generation of programming languages [56]

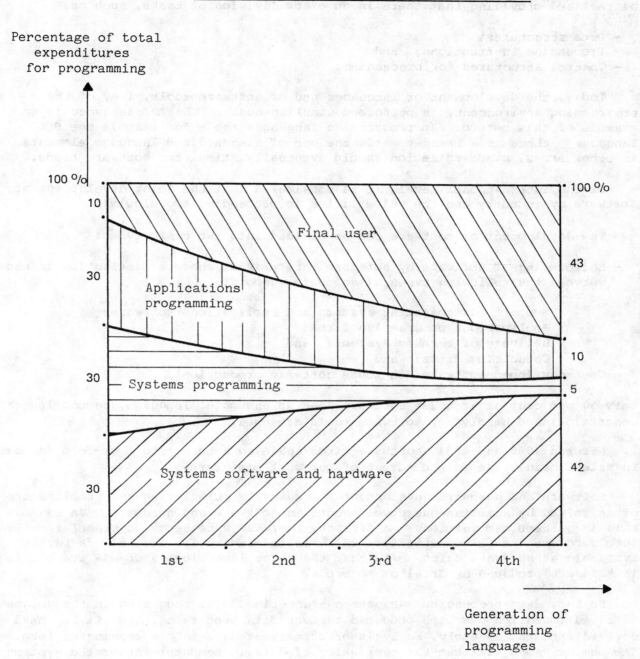

When using interactive computing techniques, it is necessary to be able to use all these elements in a parallel mode (windowing). However, this integration can only be realized providing that there is an exact division of tasks, such as:

- Data structures;
- Processing instructions; and
- Control structures for processing.

Today, the development of languages and of software tools, i.e. of the programming environment, is performed simultaneously. The ADA language is an example of this method. In programming languages too - for example the SQL language - there is a trend towards the use of standardized language elements. As in other areas, standardization should eventually stimulate software trade.

At the international level, it is considered that the joint development of software by industry and the universities is proceeding too slowly.

The development of software can be divided into two basic types:

- Engineering of software by software enterprises, where a distinction is made between the following types of software support:

        Support for operating systems and applications software;
        Applications programming firms;
        Delivery of turnkey systems; and
        Consulting firms; and
- User-developed software (in-house software production).

Only 50 per cent of in-house software uses an assembler language (second language generation) and hardly any software tools are used.

Between 1980 and 1984, software firms achieved a three- to four-fold increase in sales, mainly due to the advent of personal computers.

Software and hardware development are mutually stimulating and problems are often solved by simultaneous developments in software and hardware. An example of this is pattern recognition. At present, there is a tendency to transfer software into hardware (see also definition of firmware - annex I), e.g. MAP is partly available as hardware chips. At least the three lower levels of MAP are available in hardware produced by Intel or Motorola.

So far, data processing has been characterized by a reduction in the volume of data. Information has been produced through data processing, i.e. it has been derived from the assembly, analysis or summary of data into a meaningful form. The coming years will see the generation of data in computer-integrated systems through the automatic establishment of relations between informational objects with the help of artificial intelligence systems. Typical examples of functional and relational languages are LISP and PROLOG. These will be basic languages for systems of artifical intelligence. A LISP cell includes a pointer for the operation and one for the operand, permitting the processing of network and tree relations. LISP computers have already been developed in the United States.

III.5. Software adaptation, diagnostics, debugging, testing, verification and maintenance. Available tools

Quality assurance of software products is an area which is attracting particular attention at present. There is as yet no complex software solution in existence which is 100 per cent reliable. Exhaustive testing is often not feasible in view of the high degree of complexity of the software. The absence of semantic errors in software can only be guaranteed by automated semantic testing. The transfer from inductive to deductive software engineering is therefore inevitable.

Advantage could be gained by the automatic engineering of user-oriented software. Methods and tools of artificial intelligence are currently under development.

Of total expenditure for user developed software, up to 75 per cent may be spent on program maintenance (adaptation and further development). In practice, the trouble-shooting and editing of an error during operation cost from 20 to 100 times as much as debugging at the development stage of software. From the viewpoint of quality assurance, software products can be broken down into three types:

- Standard "off-the-shelf" operating and applications software such as personal computer software;

- Parametric software - software developed in accordance with customer specifications (for example special operating systems, text processors); and

- Software including user-oriented adaptation ("partially finished products").

Software reliability can be considerably improved by using appropriate software tools. This leads to high software engineering productivity.

A higher quality is demanded of the tools used for creating software than for the software itself. In view of the volume and complexity of current software products, software development contracts are usually negotiated with special clauses covering quality assurance and correction of operating errors.

Software is evaluated by two methods:

- Rapid prototyping for checking and testing of software drafts, especially of the user interfaces; and

- Operational evaluation by experts.

Software tools consist of the following systems:

- Text processing systems;
- Editing and file management systems;
- Comprehensive modular testing systems;
- Program linkers; and
- Simulation systems.

It is expected that, between now and 1990, with the help of advanced software tools, productivity from software design to software documentation will increase fourfold. Maintenance costs should show a reduction of up to 50 per cent.

There is as yet no software engineering method available able to support all phases of the software lifecycle from the beginning to end. Current software tools, based on relevant engineering methods, support individual phases of the software design. Since about 1983, efforts have been concentrated on the creation of data-bank based computer-aided software tools servicing all phases of the software lifecycle. The data bank constitutes the interface for interactive program development with the help of computer-aided software engineering.

Three examples of internationally accepted engineering methods are listed below:

- "Constantine" - a programe modularization method aimed at the design of a top-down structured program [90];

- SADT (Systems Analysis and Design Technique) - a method using diagrams for the description both of functions and of data [91]; and

- HIPO (Hierarchy plus Input-Process-Output) - in a joint INPUT-PROCESS-OUTPUT diagram, both data and control flows are displayed. Iterations, however, are not displayable [92].

Experience shows that, for effective software engineering, it is necessary to use a combination of various methods.

Program modularity is essential for efficient systems maintenance. Two types of modules may be distinguished for the handling of the data input/output functions:

- Functional modules; and
- Data modules.

A description of a tool for the design of manufacturing systems software for the mechanical engineering and instrument-making industries, developed in the USSR, is contained in annex VI.

III.6   Considerations of intellectual property

The non-licensed transfer of software is estimated to have caused losses of billions of dollars for software firms in recent years. Incidences of software piracy are becoming increasingly common [24].

In its model provisions for the protection of computer software, the World Intellectual Property Organization (WIPO) notes that computer programs are recognized as comprising two major elements, the underlying process, system of operation or algorithm, and the set of instructions that explain the process in detail. The underlying process, system or algorithm may be regarded as falling within the orbit of industrial property (patent) laws, while the written program itself as a set of instructions expressed in a certain form clearly falls within the scope of copyright. Both of these elements can be regarded as protectable know-how or trade secrets [3, 57]. The complex nature of computer software has caused some difficulty and confusion amongst legal experts who have attempted to analyse it exclusively in terms of patent, copyright or trade secret laws.

A group of experts on copyright aspects of the protection of computer software, convened jointly by the United Nations Educational, Scientific and Cultural Organization (UNESCO) and WIPO, has been created to evaluate the above-mentioned problems in detail.

At its meeting in February 1985, the group of experts noted the pressing need for adequate protection of computer programs, at both the national and international levels. It was stressed that such a protection should encourage software development as well as international trade in software products and that it should be based on an internationally harmonized approach [81]. Several participants expressed the view that the international copyright conventions already protected computer programs and therefore required no amendment to that effect. Other delegations expressed their doubts as to the applicability of those conventions in their present form. A great number of participants stated that computer programs were works protected by copyright, provided that they were original products, constituting individual, creative expressions of the set of instructions developed in them. The meeting noted that, during the past years, relatively few court cases had been decided and, whenever judgements had been delivered, they had by and large confirmed the applicability of copyright.

Patentability of computer programs _per se_ had been ruled out under the law of virtually every country, 1/ and other possible forms of protection under industrial property law did not grant exclusive rights to the creator of such a program. Copyright, in its development, had proved to be flexible enough to extend to works of a technical nature, such as plans.

Delegations from countries where computer programs were protected by copyright maintained that, in general, copyright provided an effective means of protection. Several delegations said that in their countries the possibility of adopting _sui generis_ protection was under consideration.

1/   In 1985, Japan's Ministry of International Trade and Industry dropped a plan to replace copyright protection of software with patent protection of shorter duration. Computer programs would have entered the public domain after only 15 years. Copyright protection, in contrast, lasts for 50 years after the author's death [82].

Copyrights are, of course, not self enforcing. The holder of a copyright must find infringing copies and their producers, and bring an appropriate legal action. Because the equipment is readily available to copy even complex programs, this is often a difficult task. Unlike printed works such as books, which may only be copied using photocopying or microfilm equipment, software programs may be copied almost automatically by the machines on which they have been designed to run. While software producers have tried various coding systems to frustrate would-be copiers, a skilful computer operator is usually successful in eventually breaking the code and copying the software [83].

IV. MEASURES TO FACILITATE THE DEVELOPMENT AND INTRODUCTION OF ADVANCED SOFTWARE FOR INDUSTRIAL AUTOMATION

IV.1. Governmental programmes launched in various countries

United States [16]

The Government's role in assisting automated manufacturing in the United States is primarily centred in R and D funding through the Department of Defense (DOD), the National Aeronautics and Space Administration (NASA), the National Science Foundation (NSF) and the National Bureau of Standards (NBS). While DOD assistance is primarily aimed at funding manufacturing technologies with specific military applications, there has been a significant shift to industry over the years.

Most of the DOD-sponsored research in manufacturing-automation technology is conducted under its Manufacturing Technology Program (MANTECH), which is funded at $200 million for the financial year 1984, of which about $56 million is dedicated to manufacturing automation. MANTECH was essentially launched in 1960 with the aims of developing and applying productivity-enhancing manufacturing technologies. It has helped develop and apply several significant technologies, including NC-machine tools and APT language for them, and calculators using integrated circuits.

There are about 400 to 500 MANTECH projects active at any time. The largest single expenditure in MANTECH ($18 million in 1983) is the United States Air Force ICAM (Integrated Computer-Aided Manufacturing) Program. ICAM has developed "architectures" for the structure and control of automated manufacturing and has funded work on CIM. In addition to the MANTECH programmes, two other DOD agencies fund longer-term, more basic research in manufacturing automation technologies. The Defence Advanced Research Projects Agency funds research in robotics, sensory control and artificial intelligence. The Office of Naval Research supports mainly university-based research in precision engineering.

NASA is currently funding research in three types of manufacturing automation, with support amounting to $5.9 million in the financial year 1984. The first of these areas is robotics and teleoperator research to develop manipulators for applications on space missions. A second area is the development of an advanced computer-aided planning systems for scheduling purposes. The third area of NASA research is in integrating CAD and CAE systems and linking them with data-management software packages.

NSF funds several programmes with application to manufacturing automation, with a budget of about $7 to $9 million in the financial year 1984. The various areas of attention include discrete manufacturing, CAD, CAM, computer-aided testing, university research centres in robotics and in material handling, touch and vision sensors, robot programming languages, computer architectures and control systems.

The NBS Center for Manufacturing Engineering conducts quite extensive research related to manufacturing automation, with a budget of about $7.5 million in the financial year 1984. Among NBS's successes has been the development of a set of standards called IGES (Initial Graphic Exchange Standards) which enables different brands of CAD systems to communicate with one another.

The Automated Manufacturing Research Facility at NBS has been developed to serve as a laboratory for various kinds of automation research. The Facility will undertake research primarily in the areas of measurement technology and interface standards.

Japan[16]

In Japan, government measures to promote the development and introduction of industrial automation, including software development, comprise direct financial assistance, R and D grants and subsidies and indirect benefits from tax policies.

Three laws have been enacted and implemented which have directly or indirectly helped the promotion of the Japanese machine-tool industry:

- Extraordinary Measures Law for the Promotion of the Machinery Industry (1956).
  This law laid the groundwork for the promotion of the machine- tool industry. It was replaced by the

- Extraordinary Measures Law for the Promotion of Specific Electronic Industries and Machinery Industries (1971).
  This law served as a basis for government promotion of rationalization/development plans for designated industries, including NC and CNC-machine tools. It was replaced by

- Extraordinary Measures Law for the Promotion of Specific Machinery and Information Industries (1978).
  In content, this law is almost identical with that of 1971, except that the 1978 law includes the promotion of the software industry.

The Japanese Government has provided financial assistance to the machine-tool industry directly in the form of subsidies and loans and indirectly through loan guarantees.

The Japanese machine-tool industry, particularly its NC-machine-tool segment, has been indirectly subsidized by the Government through its tax laws. The financial benefits/advantages include tax exemptions, special depreciation deductions and reserves. For example, in 1984, tax refunds were accorded to promote the modernization of medium and small enterprises. Under this tax law, medium and small businesses that fulfil certain conditions will benefit either from a special depreciation (30 per cent of acquisition cost) or from a special tax exemption (7 per cent of acquisition cost) when they purchase machinery and automated equipment - primarily NC-machine tools and industrial robots - that are considered necessary to make their methods of production, distribution and management more efficient and rational. If such machinery and equipment are leased instead of bought, similar preferential tax treatment is provided in accordance with leasing fees.

Government-sponsored R and D is prevalent in Japan with a high level of co-operation amongst government, industry and academy. Japan's state-funded R and D is much more closely co-ordinated and targeted than that of other market-economy countries.

One current major co-operative project is the Flexible Manufacturing System Complex, a seven-year effort to establish a manufacturing system for the rapid production of machine components in small batches. This project is being carried out by three government research institutes and 20 companies with an estimated budget of $65 million.

Direct government grants to encourage and support R and D in the machine-tool industry have been made mainly by the Japan Research Development Corporation, an arm of the Science and Technology Agency, and by MITI's (Ministry of International Trade and Industry) Agency for Industrial Science and Technology.

Since 1978, the Government of Japan has been taking specific steps to foster the development and diffusion of industrial robots. The main thrust of the government policy is aimed at co-ordinating and advancing the R and D efforts of teaching and research organizations and those of the robotics industry and, at the same time, helping accelerate the application and installation of robots throughout the Japanese manufacturing sector.

Japan's MITI takes the lead in supporting basic, long-range research, which allows private industry to concentrate its revenues on applied R and D, i.e. to the development of commercial products. Currently, MITI is conducting a seven-year $150 million research programme for the development of intelligent robots. Emphasis is placed on improving sensory perception, language systems and motion capabilities of advanced robots.

The Japan Robot Leasing Company, Ltd. (JAROL) was established in 1980 jointly by several Japanese robot manufacturers and insurance companies mainly to promote the utilization of robots by small and medium manufacturing firms with the encouragement of the Government. JAROL has expanded the Japanese market for robots manufactured by its member companies by making the robots available on leasing terms more favourable than those of private leasing companies to small and medium firms that account currently for 90 per cent of robot leases. In addition, JAROL attempts to provide engineering consultants to firms leasing robots.

The Japanese Government has provided a one-time accelerated depreciation allowance in the year of purchase for a firm installing robots between 1 April 1980 and 31 March 1983. Thus 13 per cent of the acquisition cost could be written off in the financial year 1980 or 1981, and 10 per cent in the financial year 1982, in addition to ordinary depreciation. Furthermore, a low-interest loan programme was established in 1980 to encourage the purchase of designated industrial robots by small and medium firms for use in dirty or dangerous industrial processes.

In addition to the programmes promoting industrial automation mentioned above, the Government of Japan has taken special steps to support the development of a domestic CAD/CAM software industry. It has initiated changes in copyright laws and tax policies to promote software development and issued government guarantees for this purpose. MITI is promoting a software technology centre and several information-sharing programmes.

In June 1985, the Japanese Information Technology Promotion Agency (IPA) signalled its interest in improving computer software with the announcement of a $100 million scheme - the SIGMA[1]/ Project - to run from 1985 to 1989. The objectives of this project are to produce software as a manufactured instead of a manually produced product, and to transform the labour-intensive software industry into a knowledge-intensive industry. The first step on this road is the development of a standard "Integrated Project Support Environment".

The SIGMA system consists of a centre, complete with mainframe computers, a network of over 10,000 user stations connected to the centre. Common standards run throughout the network. Software developers will have to use a standard operating system called SIGMA OS which will be based on UNIX. Users will provide the standard software development environment supplied by the SIGMA system for their own computers and build up each environment [64].

Federal Republic of Germany [62]

Much of the support for manufacturing automation provided by the Government of the Federal Republic of Germany has been made available through the medium of broadly based programmes such as the Harmonization of the Workplace Programme (Programm Humanisierung des Arbeitslebens). While support for the use of robots was not presented as the first priority of this programme, it was nevertheless an important aspect. A total of DM 700 million was spent between 1974 and 1983, of which 100 million was devoted to robotics.

The first production technology programme (Programm Fertigungstechnik) financed research and development over a range of activities - planning of construction, layout and management systems, quality, machine control, flexible manufacturing and technology transfer. Of total expenditure, 80 per cent went to small firms. While disbursements amounted to DM 163 million between 1980 and 1983 (1980 - 39.2 million ; 1981 39.5 million ; 1982 46.1 million ; 1983 38.5 million), this represented an underspending in comparison with the planned budget for the period of DM 255.6 million.

1/   S IGMA = Software Industrialized Generator and Maintenance Aids.

A further, revised programme - Fertigungstechnik was started in 1984. This programme aims at supporting the development and industrial implementation of CAD/CAM, robots and related automation systems by means of grants of up to 40 per cent of the project cost (maximum grant is DM 400,000 for CAD/CAM and 800,000 for robots and systems). Assistance is also available for co-operation projects between companies and research institutes for the development of machines and processes for FMS and inspection systems, as well as for publications, workshops, seminars and training and the study of such aspects as manpower requirements and working conditions. A total of DM 530 million is budgeted between 1984 and 1988 for the programme, with the following annual breakdown in millions of deutsche mark:

1984	1985	1986	1987	1988
70	125	130	135	70

The 1984-1988 "Fertigungstechnik" programme is part of a broader Informationstechnikprogramm for which an expenditure of DM 2,960 million is planned between 1984 and 1988. Parts of this broader programme, in addition to "Fertigungstechnik", appear to have some relevance to the fostering of automation and thus are likely to provide further support in this area.

The funds available are broken down as follows:

DM 410 million is to be spent on fibre optics and high-definition television, and the remaining DM 2,020 million as follows: computer compatibility (DM 100 million), basic research (DM 100 million), peripherals in microelectronics (DM 320 million), CAD (DM 90 million), key components (DM 90 million), sub-microelectronics technology (DM 600 million), new chips (DM 200 million), computer-assisted design methods for computer hardware and software (DM 160 million) and complex data processing (DM 200 million).

Federal support for R and D comes mainly from the Federal Ministry for Research and Technology and is chanelled through two bodies:

(i)   For fundamental research, through the Deutsche Forschungs-Gemeinschaft (DFG), which groups most universities, the Max-Planck foundation and the large scientific institutes. The DFG has an annual budget of DM 20 million for research on manufacturing automation.

(ii)   For applied research, through the Fraunhofer Gesellschaft, which groups 25 institutes. In particular, it gives support to the IPA (Institut für Produktionstechnik) at Stuttgart, which set up an experimental automated production system in 1977. IPA has an annual budget of about DM 65 million, of which some 40 million from the Federal Government and Länder sources. Another important institute that is supported by the Fraunhofer Gesellschaft is the IPK (Institut für Produktionsanlagen und Konstruktionstechnik) in Berlin (West), which also has an experimental automated system (with emphasis on testing and quality control), set up in 1976. The IS I (Institut für Systemtechnik und Innovationsforschung) is also worth mentioning: it organizes workshops and seminars dealing notably with manufacturing automation and acts as a consultant, in this area.

Other assistance that has been or still is available includes the following:

(i)   A microelectronics programme (Sonderprogramm Mikroelektronik) providing assistance, especially for small and medium industries, for the application of microelectronics. Out of the 1982-1985 budget of DM 450 million, some DM 90 million was made available to machine builders.

(ii) Provision of grants to small and medium firms for expenditure on R and D staff. A total of DM 1,407 million was spent in 1979-1982, of which about 35 per cent went to the engineering sector (i.e. DM 492 million for the period).

(iii) Provision of grants of DM 4.6 million to firms in the engineering sector for grants for external research in 1982.

Although there is little information available on this subject, the individual Länder also in some cases give considerable support to automation. In particular, the Land of Baden-Württemberg has designated a Commissioner for technology transfer, paying consultants' fees, offering technical consultancy through a network of 16 technology institutes, setting up technology factories, organizing training and supporting Karlsruhe University, which is in the process of establishing its own automated production system and has just opened the first production systems engineering course.

The preference of the Government of the Federal Republic of Germany for broadly based schemes should be noted. Within these schemes, there is a clear preference for R and D rather than direct investment encouragement, although, at a time of rapid technology change where increased R and D spending becomes an absolute necessity for companies' survival, distinctions between R and D and other types of expenditure lose much of their meaning.

Another point to be noted is the current sharp increase in support to automation. This is in fact a common feature of west European countries' public support in this field.

United Kingdom[62]

After a period during which support was given to development and investment in specific sectors (including the machine-tool - £30 million programme - textile machinery and printing-machinery industries), the United Kingdom has, since the beginning of the 1980s, tended to support the application of specific technologies and also, in some cases, the manufacture of the relevant equipment.

Thus approximately £10 million was allocated to the Robot Support Programme between 1981 and 1984 to support the production and application of robots and provide consultancies for these activities. The sum of £80 million has been allocated for 1982-1986 to the Flexible Manufacturing Systems scheme for support for the installation of FMS and for consultancies to this end. Of this amount, £25 million in fact comes from Support for Innovation Funds (see below) while £20 million is a new allocation granted in March 1984. Since 1982, the applications section of the robot support programme has been financed under the FMS scheme, and the consultancy aspects of the two schemes were merged on 1 August 1984 to form the Advanced Manufacturing Technology Scheme.

The sum of £27 million was made available for 1982-1984 under the following three schemes:

(i)   CAD/CAM. Feasibility studies, demonstration seminars and R and D projects in CAD.

(ii)   CAD/MAT (Computer-aided design, manufacture and testing). Grants for seminars and demonstrations.

(iii)   CADTES (Computer-aided design and test equipment).

The application of microprocessors to products and processes throughout the manufacturing industry is being assisted under the Microprocessor Application Project, to which £55 million was allocated for 1978-1983 and to which a further £20-30 million is expected to be allocated. A further £55 million was received for 1978-1983 for the sectoral Microelectronic Industry Support Programme with the aim of expanding the United Kingdom's ability to manufacture standard and specialized integrated circuits. The United Kingdom is also considering a £25 million scheme to foster the adoption of advanced equipment by the textile industry.

The two SEFIS (Small Engineering Firms Investment Scheme) programmes provided £130 million between 1981 and 1984 for one-third grants for small firms to buy NC-machine tools and certain other items of advanced equipment.

Grants for R and D are available from Support for Innovation Funds, which apply generally to the manufacturing industry. Some of the funds are allocated to specific programmes (e.g. parts of robot and FMS schemes), but assistance has been given to automation outside the medium of specific schemes, e.g. to the machine-tool and welding-equipment industries. The Mechanical and Electrical Engineering Requirements Board (MEERB) has also provided R and D support to the engineering sectors in general. Funding in the field of computer-aided engineering (CAE), for example, amounted to £5.1 million in 1978/79 and 2.3 million in 1980/81. This funding was granted by a predecessor of MEERB, the Mechanical Engineering and Machine Tools Requirements Board. MEERB is now an advisory body in connection with overall funding of CAE. Research into future generations of robotics is financed by the Science Research Council through a £2.5 million programme of joint industry/university research which commenced in 1980.

A major co-operative research project into information technology linking the Government, the universities, industry and other research organizations was set up in April 1983[65]. The Alvey Project's aim is to ensure the future competitiveness of the United Kingdom's IT industry by bringing together the major research efforts being made in four key areas of electronics and information technology: software engineering, very-large-scale integration, intelligent knowledge-based systems, and the man-machine interface. The five-year programme will cost £350 million and involves basic research, the establishment of a communication infrastructure and demonstrator projects aimed at bringing the results of advanced IT research to the marketplace.

The Government of the United Kingdom decided that it would supply all the funds for academic work, and that Alvey research done in industry would receive 50 per cent state funding. The cost to the Government is expected to be £200 million, 50 million of which is earmarked for academic research.

The long-term goal of the Alvey Programme's software-engineering effort is the development of the "Information System Factory", essentially a production facility for software. The projected Information System Factory will be a facility made up partly of hardware, partly of software and partly of stored knowledge. In the short term, the Alvey programme aims at promoting the more widespread use of design tools and methodologies, the development of suitable high-level programming languages, more use of formal specification methods and the limited use of automatic software-generation techniques.

The United Kingdom has general provisions for investment encouragement outside specific sectoral or technology-based schemes. It is difficult to ascertain the amount of funding involved for manufacturing automation but, for example, £4.3 million has been provided for the establishment of a Unimation robot manufacturing facility in the United Kingdom.

Thus the United Kingdom has a large number and wide variety of schemes. An increasing preference for technology-based rather than sectoral schemes is discernible, although the latter continue to exist and be proposed.

Investment assistance predominates (although frequently, for instance under the FMS scheme, investment is associated with development), but assistance also remains available for R and D. While most programmes are specific in nature, funding procedures of a very general type exist. However, it is difficult to determine the extent to which these provisions are used to fund automation.

Italy[62]

Italy does not have the profusion of support measures for automation that exist in some other countries. The Sabatini Law enacted in 1965 allowed for deferments of up to 5 years in payment of machine-tool purchases but figures for the amount of assistance involved are not available. There are three relevant programmes under the overall Progetto Finalizzato framework of the National Research Council (CNR) for which it is believed that 56,000 million Lire has been allocated:

(i)   Informatica – Lit 13,000 million was made available for work on advanced process control, CAD in mechanical engineering and machine interfacing.

(ii)   Technologie Meccaniche. This programme has a budget of Lit 30,902 million for 1983-1988. Concentrating on manufacturing technology, its main objective is the development of machine tools and their integration into flexible manufacturing systems. There are three sub-programmes: flexible systems, integrated technology and components and industrial experimentation (prototype site). From the technological point of view, the laser represents an important part of the programme, which will be managed by the CNR and involves a close liaison between research centres and industry.

(iii) Robotica. It is not known if this programme has gone beyond the feasibility-study stage.

Two general funds are of relevance. The Fund for Innovation, which supports the introduction of technological developments leading to new or improved products or processes, stood at Lit 1,600,000 million for 1981-1983. This fund applies to the automobile, electronics, iron and steel, aeronautics and chemical industries.

Law 696 of 19 December 1983 has taken over from the Sabatini Law and has now become a major vehicle for the support of manufacturing automation. This Law, under which Lit 100 million has initially been made available (taken from the Fund for Innovation), provides for a subsidy of 25 per cent of the cost (not including VAT) of buying or leasing advanced manufacturing equipment.

The IMI special fund for applied research stood at an amount of Lit 1,700,000 million for 1982/83, available in grants or loans. This fund applies to all industry.

France [62]

France has several measures offering support for automation and advanced equipment and relating both to industry sectors and to technologies.

The machine-tool plan provided for expenditures of FF 2,300 million for 1982-1985. Its aim was to restructure the machine-tool industry and make it internationally competitive, through development contracts whereby companies agreed to fulfil certain targets related to investments, R and D, capital increases, retraining, etc., in exchange for access to public funds. These, in fact, came from existing institutions and funds, such as ANVAR. The machine-tool plan also provided for co-ordination of the various public agencies involved in technological innovation and assistance for the purchase of NC-machine tools and advanced equipment by educational establishments.

There have been a series of public agencies and programmes with the aim of encouraging innovation and investment:

a)   Robotics billion - a FF 1,200 million programme providing favourable loans
     for the purchase of robots during 1982-1985.

b)   MECA procedure, which permits small and medium firms to test automated
     equipment and finances any subsequent acquisition. FF 500 million was
     allocated for 1982-85 (of which FF 150 million in 1982). The procedure is
     run by the Agence Nationale pour le Développement de la Production
     Automatisée (ADEPA).

c)   DAP (Développement de l'automatisation de la production) encouraged
     innovation in automation by small and medium companies and had a budget of
     FF 14 million in 1981 and 20 million in 1982.

d)   Efficacité des équipements et maîtrise des débouchés. The fund, which
     stood at FF 2,500 million in 1982, encouraged investment programmes
     through favourable loans of up to 70 per cent of expenditure.

e)   CODIS (Comité des Industries Stratégiques) co-ordinates the granting of
     existing public funds to selected companies in selected areas of
     technology. There is a concentration on FMS. In 1982 the fund stood at
     FF 110 million, in 1983 at 455 million and in 1984 at 715 million.

f)   PUCE (Produits utilisants des composants électroniques) has a budget of
     FF 40 million for 1983-1984 to encourage small firms to incorporate
     electronics in their products.

g)   Special loans to industry. Soft loans of FF 30-70 million to industry for
     the application of equipment in line with national priorities (including
     production automation).

     FIM (Fonds Industriel de Modernisation) was created in 1983 to provide
favourable loans to users for the modernization of manufacturing processes and
development of new products. It also took into account office technology and
biotechnology. FIM took over the activities of CODIS and possibly DAP and the
fund Efficacité des équipements et maîtrise des débouchés. Only
FF 1,000 million was spent out of FIM's 1983 allocation of FF 3,000 million.
The remainder was brought forward to 1984 as part of a total allocation of
FF 7,000 million. It has recently been reported that another FF 1,000 million
has been added to top up the fund[66].

     A programme productique for 1983-1986 has been launched; it functions
along the same lines as the machine-tool plan, i.e. through the use of
development contracts providing for privileged access to existing funds, such
as FIM, for assistance in obtaining equipment through CODIS contracts, etc. A
sum of FF 100 million has been allocated to this latter aspect.

     The following bodies provide assistance to R and D:

a)   ARA (Automatisation Robotique Avancée) co-ordinates some 50 research
     laboratories and 20 large companies for research on FMS and robotics.
     FF 5 million was budgeted for 1981 and 15 million for 1982 (another
     FF 15 million derived from another agency) and the programme was
     planned to last until 1985.

b) ANVAR (Agence Nationale pour la Valorisation de la Recherche) had a budget of FF 200 million in 1983 (out of a total budget of FF 900 million) for aids to innovation for the production of advanced equipment in the field of manufacturing technology.

Thus France has a large number of similar funding arrangements concentrating on both technologies and industry sectors of general and specific types and dedicated to users and producers. Budgets relating to these financing arrangements have not always been fully drawn on. The FIM funds were not fully spent in 1983.

As concerns major research programmes particularly intended to support software development in France, the following project should be noted:

The national software engineering project is one of six priority research projects launched as part of the Plan d'Action Filière Electronique (PAFE). The aim is to develop a universal software engineering system called EPICEA (Environnement de Programmation Industriel pour la Conception et l'Etude des Applications), by co-operation on the part of the main French companies active in this field. The public support to this project amounted to FF 30 millions in the fiscal year 1984 and the same amount is budgeted for 1985[3].

Finland[95]

The Finnish Programme for Research and Development in Information Technologies (FINPRIT) was initiated in 1983 by the Finnish Ministry of Trade and Industry with the participation of Finnish industry. R & D was started in June 1984 and it will continue until 1988.

The total budget of the programme is 100 million FIM (21 million ECU's, 19 million USD). About 70 per cent of the budget is covered by the technology Development Centre in Finland and most of the rest by 30-40 Finnish companies participating in the programme.

The level of work is mainly applied technical research and on a concrete level, the results are typically prototypes and demonstration systems. The ownership rights of the results belong to the research institutes, but the participating companies have the right to use the results in a specified way.

The main goal of the FINPRIT programme is to improve the international competitiveness of Finnish industry in the area of information technology. The programme has five subprogrammes, each of them consisting of projects, 21 in all. Each project has a Technical Support Group representing the research institutes and the companies, and each subprogramme has an Executive Committee. There is also a Programme Board to stimulate co-operation within the programme.

The main research organizations are the Technical Research Centre of Finland, Helsinki University of Technology, the University of Oulu, Tampere University of Technology, and the University of Tampere.

The five sub-programmes of the FINPRIT programme, mentioned above, include the following:

- Integrated office systems

The integration of telecommunications, data base services and personal data processing is based on offical and de facto standards. Prototypes to be developed will be general and portable. Specialized software tools are also to be developed, the ultimate goal being a demonstration system to test the integration techniques.

- Distributed CAD/CAM systems

The growth of automation technology in the production and integration of information systems is changing industrial manufacturing methods and the information systems supporting them. Efficient tools for product design and production management are the key factors in industrial competence. Advanced software tool prototypes are being developed especially for distributed CAD/CAM. Powerful workstation computers, UNIX and local area networks are being used.

- Industrial software engineering

The amount of software in embedded systems is constantly increasing. The improvement of software development productivity in this area has proved to be a difficult task. As a rule embedded systems are hard to standardize, and useful application generators are rare. The subprogramme philosophy is in harmony with the general world-wide trends in software engineering.

- Pattern recognition

The rapid development of micro-electronics has enabled the economic application of methods of pattern recognition and image processing. The aim of this subprogramme is to develop and acquire know-how on the industrial applications of pattern recognition. This technique would appear to be a key factor of competition in several areas of application.

- Expert systems

Artificial intelligence and especially expert systems have succeeded in getting into the commercial level. This subprogramme has a very pragmatic attitude towards the development of concrete operational expert systems using the state of the art hardware and software tools.

In Finland, apart from FINPRIT, many other related governmental R and D programmes are also proceeding currently, for example:

- Microelectronics (specified fields e.g. IC-production);

- Transducer technology;

- Factory automation;

- Laser technology in heavy engineering; etc.

Netherlands 1/

In the field of software, stimulating measures have been taken in the Netherlands. A particular measure, is the Stimulation Scheme for Computer Service Industries, which includes six software application fields, computer integrated manufacturing being an important one.

Until the end of 1986, proposals could be submitted for the development of new software. Grants of up to 40 per cent are approved if the software is: applicable by several clients, user-friendly, and compatible with national and international standards. For the total scheme, a budget of 43 million Dutch guilders is available.

Czechoslovakia[63]

Of the government programmes existing in Czechoslovakia in the field of electronics and automation, the following three are the most important:

- The advancement of microelectronics;
- The development and application of automation means; and
- The automation of engineering production processes (discrete manufacturing).

During recent years, independent programmes have been formulated for robotics and the application of industrial robots.

The common characteristics of these programmes are the emphasis on R and D and on the following aspects:

- Identification of problems and development phases;
- Determination of steps in the development of equipment including its parametrization and the definition of its functions;
- Development of elements, functional groups and assemblies;
- Development of computers and peripherals with standardized interfaces;
- Development of corresponding software;
- Application of the equipment in practice and optimization of operation conditions;
- Development of working places provided with the equipment in question;
- Training of specialists.

1/ Transmitted by the Government of the Netherlands.

The programmes also have certain specific features, connected partly with corresponding development phases and partly with fields of implementation and influencing their internal structuring in certain cases. The majority of these programmes run for five years, in conformity with the Czechoslovak five-year economic plan structure.

The above-mentioned governmental technical programmes cover the following areas:

Microelectronics

In this highly progressive area, the programmes are aimed in two directions: R and D of components; and relevant applications. The first comprises programmes of R and D in specific areas, their integrated application, miniaturization, parameters and function testing, their specific industrial production technologies and the development of special machines. In particular cases, it may also cover the development of specific equipment. The second examines the use of microelectronics in particular branches of the economy. Emphasis is on the application of electronics in all areas, at present in particular in the construction of industrial machines and equipment.

Automation means

Programmes in this area cover, in the first place, the development of automation components and corresponding means, especially computer systems and computer networks. The other programme direction is related to the development of application software for particular areas. The standardized solution of Management Information Systems (MIS) resulting from this programme has been implemented in several hundreds of Czechoslovak industrial plants. Another very important programme is the computer automation of technological process control. This programme also involves the development of hierarchical networks, which will be the main orientation of an upcoming programme.

A new programme oriented towards CAD/CAM systems was recently added to the above programmes, while the programme dealing with robotics was moved from this group into another one.

Automation of engineering production processes

The programme oriented towards the appliation of automation in discrete manufacturing has been of prime importance in the current decade. In view of the dominating position of mechanical and electric engineering in Czechoslovak industry, this branch plays a vital role. Research and implementation work is directed mainly to the application of flexible automation. Upon conclusion of the programme for the introduction of production methods and the implementation of NC-machines in the first half of the 1970s, the programme for the first flexible manufacturing systems (FMS) with a lower level of automation was launched. This was followed, in the first half of the 1980s, by the programme for FMS with a higher level of automation. The current programme is aimed at the application of FMS in a group mode and the application of flexible automation on the shop floor and in the whole factory.

Industrial robotics

The first programme in this area was based on the results of basic research and led to the establishment of industrial robots for manipulation and the performance of certain specific operations. The current programme deals with the next generation of industrial robots - adaptive robots - and covers the development and application of robotized complexes and working places and the R and D of sensors. This programme will also deal with the application of artificial intelligence.

Responsibility for the programmes

The above programmes are under the main control of the State Commission for Technical Development and Investments. The responsibility and co-ordination functions have, in the case of some programmes, been transferred to other governmental bodies - in some cases branches of ministries (for example mechanical engineering, electronics, energy, etc.). In all cases, the selected governmental research institute becomes the executive body of the programme and plays the central role at every step of development. The governmental bodies are able to follow the progress continuously as regards both its content and financing and the results achieved.

Hungary[2]

The concept of integrated data and material processing systems was first studied in Hungary in 1971 under the auspices of the State Committee for Technical Development (SCTD). As a result, a three-year research programme was financed. On the basis of its final report, the SCTD called upon experts from industry, university and academe to set the goals for the introduction of a few systems in Hungarian industry. On the basis of their recommendations, five manufacturing systems were designed, developed and installed with considerable state support.

From experience gained during test runs and subsequent production operation, it became obvious that the integration of data and material processing systems required new design methodologies, since the development of the components in the system was much more rapid than the development of synthesis theory. In 1976, the SCTD established a committee to study the world-wide tendencies of such systems and to define manufacturing models for the 1980s. The committee concluded that the model for the next few years would differ considerably from the long-range model.

The entire proposal of the committee was circulated in both academic and industrial circles and the SCTD called together competent professionals for plenary discussion. Corrections were made and the final research proposal and budget were adopted.

Table 6 lists the specific research topics that should be covered in the project. Although the specifications of the research programme emphasized short-range problems, they also required that long-range research tasks should be analysed.

Table 6.   Topics to be covered in the study of integrated data
and material processing systems in Hungary

Short-range problems	Long-range problems	General topics
System configurations and criteria System design methodologies Interfaces General monitor system	Systems engineering	Analysis and synthesis of integrated data and material processing systems
Databases Production planning systems	Production control systems	Integrated production control, information and decision making systems
Technological planning	CAD/CAM interface	Computer-aided engineering
Completion of existing manufacturing systems	New control systems	Selected modules of computer-integrated manufacturing
Adaptive control	Optimization and adaptive control Learning procedures Computer-controlled robots Diagnosis and maintenance	Application of artificial intelligence and optimization methods

IV.2 International co-operation

IV.2.1 Programmes set up by intergovernmental organizations

European Strategic Programme for Research and Development in Information Technologies (ESPRIT) of the European Economic Community (EEC) [67]

In 1983, the member states of the European Community agreed on a long-term programme for joint research and development efforts in information technologies (ESPRIT). The strategic objective was to be the achievement of technological parity with, if not superiority over, world competitors within 10 years. The financial resources for the programme, which would be implemented in two phases (1984-1988 and 1989-1993), would be mobilized for the first phase in an amount of some 1,500 million ECU of which EEC would have to provide 50 per cent.

The research undertaken in ESPRIT is termed precompetitive, i.e. the focus is on basic technology, not marketable products. Those are left up to each partner. The technological domains selected to be subject of the programme included the following five:

(a) Advanced microelectronics capability;
(b) Software technologies;
(c) Advanced information processing;
(d) Office systems; and
(e) Computer integrated manufacture.

As concerns software technologies, ESPRIT expects that in the next decade software costs will rise to make up 90 per cent of the overall system development cost of information technology systems. The importance of early standardization and co-ordinated action in the software development field is stressed. Software development is recognized as an intellectual and industrial, as well as a commercial and entrepreneurial activity. On the basis of this combined approach to the software development process, ESPRIT defines three main research themes:

(a) Theories and methods for programme development;
(b) Methods and tools in software engineering; and
(c) Economics of industrial software production.

The reason for selecting computer-integrated manufacturing (CIM) for an ESPRIT domain is explained by expected significant market prospects and the foreseen positive impact on manufacturing productivity. The main areas of R and D activities in the CIM domain in ESPRIT are:

(a) The integrated system architecture area, covering:

- Identification and development of overall integrated system structures for data-base systems for engineering data of total product models and manufacturing data of plant, machines and tool models,

- Data-base management systems aimed at ensuring the required data communication between the components of the integrated manufacturing system and the data bases.

(b) The system and general software area, covering:

- Computer-aided design/computer-aided engineering systems aimed at an improved design process in respect of both shorter design time and accuracy and at establishing total product models for subsequent use in various stages of the manufacturing process,

- Computer-aided manufacturing systems aimed at formulating modular CAM system structures allowing for all types of applications in all sectors of industry,

- Computer-aided testing/computer-aided repair aimed at cost-effective improvement of product quality,

- Command languages aimed at developing software modules capable of generating control programmes from design/production/test simulation data for robot manipulators, computer numerically controlled machine tools and flexible manufacturing systems.

(c) The machine control area covering:

- Automated assembly and assembly operating systems aimed at establishing fully automatic assembly systems,

- Robot operating systems, where future areas for robot applications will pose requirements different from those being met today,

- Imaging (global and control), where future systems will require the use of complex imaging "sensory input" for CIM applications in such areas as assembling, machining, testing, etc.

- Computer numerically controlled machine tools (CNC-machines), where new application areas within metalforming and other mechanical engineering industries are expected to develop.

(d) The component area, covering:

- Sensors, where progress is considered necessary for the future development of advanced automated manufacturing systems,

- Microelectronic sub-systems aimed at integrating entire control sub-systems on to single chips.

One of the projects accepted under the CIM scheme of ESPRIT is a joint effort by COMAU (Italy), Renault Automation (France) and Digital Equipment Corp. (through its subsidiairy in the Federal Republic of Germany) to design and develop a series of software modules for CIM, especially for small-batch manufacturing [68]. The project, involving an estimated expenditure of more than $10 million, will focus on production-activity control to optimize manufacturing productivity. The goal of the project is to design, develop and test software modules required to complete the automation of the production process and to reduce human intervention and reaction times as much as possible, relying on data automatically captured from the plant.

Nine west European commercial concerns (ICL, the Atomic Energy Authority and STC of the United Kingdom, Informatique Internationale and Thomson of France, AEG and GRS of the Federal Republic of Germany, Italy's Esacontrol and Elektronikcentralen of Denmark) are working together on another project to improve the reliability of software, in particular in the computers that operate factory equipment. The companies will produce techniques to monitor software reliability and attempt to amass information about programme failures [69].

The first years of ESPRIT have proved that setting up research projects involving universities and companies from different countries takes time. Much of the time was spent setting the ground rules for co-operation and becoming acquainted with other partners. Nevertheless, the first ESPRIT patent has been issued, covering advanced interconnection for VLSI circuits; a demonstration signal processing chip has been fabricated; and a book of design rules for automated manufacturing has been published [70].

Recently, an interim report was released about some results of the first two years of ESPRIT. It is stated that about 90 per cent of the funds for the first phase (1984-1988) are already allocated or promised. The report underlines the stimulating effect of ESPRIT on west European co-operation in research on information technology. Some key figures for ESPRIT in 1984 and 1985 are given below in table 7.

Table 7: ESPRIT in 1984 and 1985

	1984	1985 a/
Project proposals submitted	441	389
of which: accepted	110	95 b/
Project participants	554	565
of which: Firms	326	316
Universities	107	121
Research Institutes	95	70
Other	26	58

Source: [71]

a/ Provisional figures
b/ In part not yet passed.

EUREKA[1]/

The west European action for co-operation in research and development - EUREKA - was created by 17 countries and members of the Commission of European Communities in Paris on 17 July 1985. The basic idea of this project was to give fresh impetus to western Europe's efforts to keep technologically abreast of the United States and Japan. It was generally recognized that governmental aid would be needed for certain projects, particularly those of a longer-term and riskier nature.

The detailed goals of the Eureka scheme were agreed upon by a second ministerial conference of 18 countries and the EEC at Hannover (Federal Republic of Germany) on 5 and 6 November 1985. The "Declaration of Principles", adopted by the meeting, sees EUREKA as a framework within which companies, generally the prime movers of individual projects, can co-operate across borders, assisted by diminishing national barriers - and ultimately a genuine internal common market, to which the European Community is committed by 1992 [72].

EUREKA covers mainly the following areas of R and D: information - and telecommunication technology, robotics, new materials, manufacturing technology, biotechnology, marine technology, lasers and environmental and transport technologies.

A group of 10 multinational research projects was formally adopted, involving companies from 12 nations, as well as the EEC. These co-operation projects include:

(a) Creating a west European standard for personal and educational microcomputers. Participants: Olivetti (Italy), Acorn (United Kingdom), Thomson (France).

(b) Compact vector computers for high-speed calculations. Participants: Norsk Data (Norway), Matra (France).

(c) Flexible manufacturing based on opto-electronics. Participants: CGE (France), Comau/Fiat (Italy), LASG (Switzerland).

(d) European Research Network. Participants: Federal Republic of Germany, France, Finland, Netherlands, Austria, Sweden, Switzerland, EEC.

During another meeting, in London on 22-23 January 1986, the representatives of the 18 west European countries participating in the EUREKA project defined the form and functions of the EUREKA Secretariat and gave the go ahead for 16 new projects [73].

---

[1]/   EUREKA is a modified acronym for European Research Coordination Agency.

The EUREKA Secretariat will be composed of six members, three will be nominated by the EEC countries, two by non-EEC countries and one by the Commission of European Communities. It will be managed by one of the six members, who might well be a high-level official or an industrialist. The Secretariat will be assisted by a team of six people. Its first job will be the development of a data bank which will allow permanent links to be established between participating countries.

The high-level officials also finalized the procedure for attributing the EUREKA "label". Under this procedure, no more than 45 days will be needed for a project to obtain this recognition, the Secretariat acting as mediator between the institutes or industrialists, the Governments and the west European institutions involved. The "label" will only be attributed to projects submitted by firms situated within the 18 countries concerned and whose work will benefit these countries.

Finally, the officials approved 16 new corporation projects to add to the ten accepted at Hannover in November 1985. The 16 new projects concern pharmaceutical products, telecommunications, environment technology, synthetic images, integrated circuits, robots, lasers and information technologies. These co-operation projects include:

(a) Creation of a flexible automated factory for the production of electronic equipment. Participants: Eurosoft S.A. (France), Sagem (France), CSEA (Italy), Inisel (Sweden).

(b) Creation of a European centre for new image synthesis technologies. Participants: Sesa (France), RTL Productions (Luxembourg), Barco Creative Systems (Belgium).

(c) EUREKA Advanced Software Technology: Development of software workshops that include software engineering. Participants: SFGL (France), CRI (Denmark), Nokia (Finland), Datamat (Italy), Intecs (Italy), Sesa Italiana (Italy), Selenin (Italy), CIR (Switzerland).

(d) Creation of an automated production management system using developments in artificial intelligence. Participants: Aerospatiale (France), Seri Renault (France), Bull (France), Cert (France), Brown Boweri (Switzerland), Aeritalia (Italy), Matrici (Sweden), Ikoss (Greece), Absy (Belgium).

Finally, at their meeting in London on 30 June 1986, ministers from the 18 countries participating in EUREKA gave final approval to 62 pojects (including the 16 projects already provisionally adopted at the previous meeting in London in January 1986), involving total spending of about $US 2.1 billion [80]. The projects mostly cover computers, semiconductors, software and telecommunications. New projects include:

(a) European Software Factory: Design and creation of a data base with programming modules accessible to firms engaged in software development. Participants: France, Federal Republic of Germany, Norway, Spain, Sweden.

(b) Prolog Tools: Development of software tools in the programming language Prolog for the creation of expert systems. Participants: Belgium, Federal Republic of Germany, Switzerland.

(c) Field bus: Communication architecture based on LANs for real-time control of industrial processes and machines. Participants: Finland, France, Italy, Portugal and United Kingdom.

(d) BD 11: Development of a data base for distributed expert systems on low-level computers, using the Pick operating system and "C" language. Participants: France, Spain.

The ministers also agreed to admit Iceland as EUREKA's nineteenth member, to establish a small EUREKA secretariat in Brussels which will act mainly as a clearing-house for projects, and to study further ways of obtaining private venture capital for EUREKA projects.

<u>Comprehensive Programme to Promote the Scientific and Technological Progress of the Member Countries of the Council for Mutual Economic Assistance (CMEA) up to the Year 2000</u>

The forty-first special meeting of the Session of the Council for Mutual Economic Assitance (CMEA), which took place in Moscow on 17-18 December 1985, adopted the Comprehensive Programme to Promote the Scientific and Technological Progress of the Member Countries of CMEA Up to the Year 2000 (hereafter referred to as the Programme).

The CMEA member countries, bearing in mind that during the next five-year period many research institutes, research and production combines and industrial enterprises would be involved in development and production on the basis of specialization and co-operation in priority areas, including the establishing of direct links, agreed to take all the necessary steps to create the organizational, legal, economic and other conditions for the implementation of the Programme in good time. The Session instructed the organs of CMEA and the international economic organizations of the CMEA member countries to make the Programme the basis for their activities.

The Programme is intended to help to promote mutual co-operation, collaboration and specialization in production and to strengthen the material and technical basis of the CMEA countries.

The CMEA member countries have agreed on concerted action to develop and exploit distinctly new types of technology by focusing their efforts and organizing closely integrated co-operation within the CMEA framework in five priority areas:

- Computerization of the national economy;
- Comprehensive automation;
- Nuclear power;
- New materials, and the technologies for producing and processing them;
- Biotechnology.

- 136 -

The goals of co-operation among CMEA countries in the field of comprehensive automation include the development and introduction of flexible automated production units, rotary conveyor belts, industrial robots, automated equipment with built-in control systems, precision equipment, high-precision meters, automated means of controlling technological processes and equipment, especially precision equipment, the ensuring of their reliability, and extensive use of integrated systems.

To attain this aim, it is intended, as a first step, to develop:

- Quickly readjustable and flexible manufacturing systems for various purposes, as well as fully automated workshops and plants;

- Systems for automated design and tooling up for production; automated and accelerated means of research and testing; automated systems of controlling production and technological processes; and integrated control systems;

- Industrial robots and manipulators for various sectors of the national economy, including seeing robots that obey vocal commands and are programmed and quickly adjustable to changing working conditions;

- Automated technologies for manufacturing ultra-high-precision equipment and instruments;

- Standardized accessories: mechanical, hydraulic, pneumatic, electrical, electronic and other components; a number of advanced control and diagnostic devices for automated machines and technological equipment; on-line control systems; highly durable instruments and attachments; and quality-control systems;

- Standardized sets of automated devices for lifting/conveying, cargo-handling and conveying/stacking work, waste-utilization systems, and sensors for various uses in industrial robots and flexible manufacturing modules.

The Programme is intended to reduce the cost of design and manufacture by approximatey one third, ensure wide interchangeability of units and modules manufactured in the CMEA countries, halve the labour inputs needed to produce them, and boost labour productivity of lifting/conveying, cargo-handling and conveying/stacking work by at least a factor of four.

Development and large-scale use of systems for automated design and control of technological processes and production will make it possible to improve planning, accounting, control and organization of production and to reduce the lead time by one third or one half.

The development of flexible automated manufacturing systems and their wide introduction in the national economies of the CMEA member countries is intended markedly to improve production efficiency and to make it possible to reduce the time and cost involved in turning out new types of products by one third or one half, raise labour productivity by a factor of two to five, bring the ratio of shift-utilization of equipment to 2.8, cut the numerical strength of operatives, and establish conditions for creative and attractive work.

The attainment of these and certain other objectives within the priority area entitled "Comprehensive automation" will profoundly influence all spheres of life and work in society, will result in a major rise in labour productivity in the basic sectors of the economy, especially the machine-tool industry, and in product reliability, quality and competitiveness, boost capital productivity, drastically reduce manual and low-skilled labour, and substantially improve the general technological level and efficiency of production.

The CMEA member countries agreed that the obligations they had assumed under the Programme over the next five-year period would be included in their plans for 1986-1990. Subsequently, amendments and additions will be periodically made to the Programme on the basis of their proposals, taking into account social, economic, scientific and technological progress made, new world achievements in science, technology and industry, and topical aspects of co-operation among CMEA members. The Programme will be developed and updated to take account of the relevant agreed measures on co-operation in the course of co-ordinating economic development plans and in national five-year plans.

The CMEA members will pay special attention to providing the necessary material and financial resources for the co-operation measures envisaged in the Programme.

The priority tasks will be organized under the leadership of the main co-ordinating organizations, which will be responsible for high technical standards and quality, for the fulfilment of projects on schedule and also for the drafting of agreements, treaties and other proposals on co-operation to be submitted for consideration by the participating countries in accordance with the established procedure.

The CMEA member countries will promote direct links between their enterprises, combines, and scientific and technical organizations based on the provisions of specific bilateral and multilateral agreements and treaties. This is an effective means of promoting co-operation for the implementation of the Programme.

With a view to the joint solution of major problems, the countries involved will, if necessary, set up joint research, development and production combines, international engineering and technology centres for the development and production of new technologies and materials, centres for teaching, training and advance training of personnel, and other joint enterprises and international science and technology teams and laboratories.

The agreed co-operation measures for carrying out the Programme will be financed by means of national funds, loans from the International Investment Bank and the International Bank for Economic Co-operation, and also out of joint funds set up by interested countries to finance major individual projects to be specified in the agreements and treaties.

CMEA member countries which do not participate in projects envisaged in the Programme are entitled to receive the results of research and development on conditions agreed upon with the countries performing such work.

More attention needs to be paid to the provision of personnel for the Programme. Accordingly, interested countries will draw up and put into practice agreed measures to promote co-operation in personnel training and retraining.

The CMEA member countries will work together in the field of technical standard-setting and measurement in the priority areas of the Programme. This will include elaboration of the necessary CMEA standards, rules, methods and means of measurement corresponding to the world's highest levels.

Each member country of CMEA is entitled at any time to express its interest in participating in individual co-operation projects under the Programme, on conditions agreed upon with the participating countries.

The Council will regularly consider major questions of economic, scientific and technical co-operation relating to the implementation of the Programme and will determine the main trends of its future development.

During the forty-first special meeting, the participants, in the context of the implementation of the Programme, signed three agreements, namely: the General Agreement on Multilateral Co-operation in the Development and Introduction of Automated Planning Systems, the General Agreement on Multilateral Co-operation in the Development, Production and Use of a Common System for Optical Data Transmission, and the Agreement on the setting up of INTERROBOT, the International Research and Production Combine for the Development of Robotics.

IV.2.2 Activities of non-governmental organizations

International Federation for Information Processing (IFIP) [75]

Advancing the state of the art in various facets of computing is one of the aims of the International Federation for Information Processing (IFIP), a federation of national professional bodies. IFIP's technical work is organized by a number of Technical Committees (TCs), each of which delegates its professional work to specialized working groups (WGs). The aim of TC 5 is to promote and co-ordinate the exchange of information on computer applications in technology. Its scope includes all aspects of computer applications in technology, i.e., in the research, design, manufacture, operation and control of products and physical systems; it also includes the investigation of related programming methods. The work in TC 5 is carried out in the framework of six working groups: WG 5.2 (Computer-aided design), WG 5.3 (Discrete manufacturing), WG 5.4 (Common and/or standardized hardware/software techniques), WG 5.6 (Maritime industries), WG 5.7 (Automation of production planning and control), and WG 5.8 (Product specification and product documentation).

Three significant topics will influence future work of WG 5.2 (Computer-aided design): Human/computer interaction, technically oriented data bases, and expert systems.

WG 5.3 (Discrete manufacturing) deals, in particular, with robotics. Key problems include the combination of high precision of computer-controlled positioning with the ability to lift heavy loads, and advanced vision and force-sensing systems. A triennial international conference on computer software for discrete manufacturing (PROLAMAT) is sponsored jointly by WG 5.3, AFCET[1] and IFAC.[2]

WG 5.4 (Common and/or standardized hardware/software techniques) operates also as the European Workshop on Industrial Computer Systems (EWICS). It provides a working body for the preparation of pre-standardization proposals, and for monitoring and participating in international standardization activities. It aims at promoting the efficient use of industrial computer systems through education, dissemination of information and the development of standards and guidelines.

The scope of WG 5.6 (Maritime industries) includes methods of automation of the operations of shipping companies and shipyards, of oceangoing vessels and smaller craft, of fisheries, and of other maritime industries and installations, topics in the use of the associated computers, signal-gathering equipment, data-presentation devices, and associated equipment and techniques.

WG 5.7 deals with various aspects of the automation of production planning and control. Important topics covered are the modelling of production management systems, and a closer integration between production management systems and those for design and manufacture.

The formation of WG 5.8 (Product specification and product documentation) reflects the implications of the trend for products to be specified not by drawings but in numerical form, and for this to be followed through at all stages of the manufacturing sequence. Thus, an immportant sub-area consists of the interfaces between product specification and CAD, between CAD and product documentation; and between product documentation and CAM and production management.

International Federation of Automatic Control (IFAC) [76]

The International Federation of Automatic Control (IFAC) is a multinational non-governmental federation of national member organizations, each one representing the engineering and scientific societies concerned with automatic control in its own country.

1/ Association Française pour la Cybernétique Economique et Technique.

2/ International Federation of Automatic Control.

IFAC provides a framework for collaboration between those working in automatic control and systems engineering. Technical work of IFAC is carried out in 14 Technical Committees (TCs), some of wich have formed Working Groups (WGs). The following TCs are of particualar relevance in respect to the subject of the present study: TC on Computers, TC on Manufacturing Technology, and TC on Economic and Management Systems.

The Committee on Computers is concerned with the utilization of computers in the control of both continuous processes and discrete manufacturing. Technical areas of interest include software, safety and reliability, hardware, distributed control, interprocessor communication, interfacing and data-base management. At the same time, the Committee serves as a major link between IFAC and IFIP.[1]/

The Manufacturing Technology Committee deals with the information-control issue associated with manufacturing processes. Illustrative processes are: material processing, material handling, inspection, assembly, robotics, process oriented computer-aided design andtechnological aspects of management information systems. Technical areas of interest include mechanics, motion control, artificial intelligence and pattern recognition, information systems, associated computer and software systems, synthesis and design technique, man-machine systems and microeconomic modelling.

The Economic and Management Systems Committee is concerned in general with modelling, analysis, control and management of large systems in which self-organizing behaviour can be present from the point of view of economics, finance, organization, resource allocation and decisions. The development of dynamic planning and management processes and the interactions between the various systems involved, in the face of external exchanges, are of particular relevance. The first area of application is that of macroeconomic national, regional, sectoral or global modelling and control problems. The second area is management planning and control in industrial and public organizations, with scheduling, investments, operations and manufacturing. The third area is hardware, software and orgware integration in relation to management information systems, innovation, man-machine interactions and general effects. Finally, to sustain these developments, the Committee is involved in software tools, forecasting, estimation and control methods to the extent of their application and use in the above areas only.

International Organization for Standardization [77]

The International Organization for Standardization (ISO), the specialized international agency for standardization, at present comprises the national standards bodies of 89 countries.

_____

1/ International Federation for Information Processing.

- 141 -

The scope of ISO covers standardization in all fields except electrical and electronic engineering, which are the responsibility of the International Electrotechncial Commmission (IEC). The results of ISO work is published in the form of International Standards.

The technical work of ISO is carried out by about 2,400 technical bodies (of which 164 technical committees, 645 sub-committees and more than 1,500 working groups). Of particular interest in respect to the subject of the present study are the activities of TC 184: Industrial automation systems. TC 184 was founded in 1983 and its scope covers standardization in the field of industrial automation systems encompassing the application of multiple technologies, i.e. information systems, machines and equipment, and telecommuncations. The work in TC 184 is carried out in 5 sub-committees (SCs): on NC-machine tools (SC 1), industrial robots (SC 2), data communications and networks (SC 3), data representation (SC 4) and systems integration (SC 5).

In this framework, SC 5 is recognized as a planning and assistance body for all standardization activities in the area of industrial automation systems, i.e. it serves as a co-ordination body for the whole TC 184. The area of work of SC 5 has been defined as the identification of requirements for new standards and the development and definition of reference models for system integration in the area of industrial automation in a manufacturing environment, from product conception through distribution.

For careful consideration of all problems related to information procesing in general, TC 184 has established liaison, <u>inter alia</u>, with TC 97 "Information processing systems". An important recent result of the work of TC 97 as regards the subject of the present study was the approval of International Standard 7498 "Information processing systems - Open Systems Inerconnection - Basic Reference Model" in 1984.

The growing complexity of problems related to industrial automation and the application of information technology has called for increased and strengthened co-operaton between ISO and IEC. This recently led to the establishment of two Joint Groups:

(i)   The Joint Group on Industrial Automation

      This Joint Group co-ordinates the activities of ISO's TC 184 with those of IEC's TC 44 (Electrical equipment for industrial machinery) and TC 65 (Industrial-process measurement and control).

(ii) The Joint Group on Information Technology Mangement

      This Joint Group co-ordinates the activities of ISO's TC 97 with those of IEC's TC 47B (Microprocessor systems) and TC 83 (Information technology equipment).[1]

1/   See also under following heading: International Electrotechnical Commission.

International Electrotechnical Commission (IEC) [78, 79]

The International Electrotechnical Commission (IEC) is the authority for world standards for electrical and electronic engineering. The Commission is composed of National Committees from 42 countries. The IEC works in close co-operation with many international organizations including ISO, which is responsible for international standards in the non-electrical fields. IEC world standards are prepared by more than 200 specialized committees, on which all National Committees have the right to be represented, and some 700 working groups. They are adapted and published by consensus of the National Committees.

In industry, the move to automation produced demands for IEC standards for electrical and electronics equipment for industrial machines. The introduction of control systems into industrial processes gave rise to the need for IEC standards, first for analogue systems and later for digital data communication systems. To have access to information, to gather, process, transfer and exchange it means that this equipment must be able to interconnect as part of a system, and communicate using the same protocols or rules. IEC work here ranges from buses to link microprocessors, to buses to interconnect instruments and buses for complete industrial process control systems. Work is nearing completion on two microprocessor system buses which will allow data transfer between a variety of microprocessors. At the instrument level, a standard for an interface system for programmable measuring instruments has been issued. Work on "PROWAY", a system of process data highways to control industrial process equipment, is well advanced. IEC prepares software standards for certain electronic systems. These are based on ISO standards which cover basic aspects of software and there is continuing co-operation between ISO and IEC.[1]/

The work carried out by the following IEC Technical Committees (TCs) is of particular relevance to the subject of the present study: TC 44 (Electrical equipment of industrial machinery), TC 65 (Industrial-process measurement and control) and TC 47 (Semiconductor devices), in particular Sub-Committee 47 B (Microprocessor systems), and TC 83 (Information technology equipment). The work of TC 56 (Reliability and maintainability) also icludes software aspects.

Subjects relating to software currently under consideration in the above Committees include the following:

(i) Microprocessor systems - extension of high level languages (TC 47 B/WG 2);

(ii) Microprocessor systems - extension of high level languages (TC 47 B/WG 2);

1/ See also under previous heading: International Organization for Standardization.

(iii) Guidelines on assessing the integrity of the design and related
      aspects of systems which contain both hardware and software
      (TC 56/WG 10); and

(iv) Preparation of relevant international standards on the subject of
     safe and reliable software (TC 65A/WG 9).

International Federation of Robotics (IFR) [96]

In autumn 1986, the International Federation of Robotics (IFR) was
established. This organization will promote development, use and
international co-operation in the field of robotics. It will also serve as a
focal point for information about robotics development throughout the world.
Membership in IFR is open to all countries. Countries which have already
confirmed their intention of joining IFR are: Australia, Bulgaria,
Czechoslovakia, Federal Republic of Germany, France, German Democratic
Republic, Italy, Japan, the Netherlands, Poland, Spain, Sweden, Switzerland,
the United Kingdom and the United States. Several additional countries are
expected to join IFR shortly, including many Asian and third-world countries
that are just beginning to utilize robotics.

V. CONCLUSIONS

There is no doubt that software is a very dynamic component of industrial automation processes.   Specific tools are needed for its development and application.   While most of the basic and user software is produced by computer firms and software houses and organizations, adaptation, maintenance and further modification of the programs give rise to significant additional heavy expenditure on the part of the user.

Reliable software solutions are available in many branches of flexible automation in discrete manufacturing.   This applies in particular to:

- Programming of numerically controlled machine tools;

- Computer-aided production planning;

- Programming of computerized numerical control (CNC);

- Realization of direct numerical control (DNC) modes; and

- Control of flexible manufacturing systems (FMS).

Current emphasis is on the development of software solutions in connection with the increased application of industrial automation, in particular, industrial robots and handling, transport and storage equipment.   Special attention should be paid to developments related to external (off-line) programming as this method not only makes possible the delivery of reliable control information for high-quality expensive automation equipment, but also the avoidance of significant down times.

The application of this equipment together with the increasing trend towards low-attended manufacturing are giving rise to growing demands for sophisticated software packages for process control, supervision and diagnostics.   This software must be equipped to deal with the input from sensing, control and monitoring devices for the realization of special modes (for example ACC, ACO).   The supervision function in the control of automated manufacturing equipment can take the form of a hierarchy comprising the following levels:

- Status detection;

- Status supervision;

- Adaptive control; and

- Self-learning control.

Diagnostic functions can be integrated into each level.

Initially based on the development of electronic data processing and man-machine communcation techniques, the automation of engineering decision-making has rapidly attained a high level (CAD, CAE, CAM etc.).   It should be emphasized that there is no branch of engineering which will remain untouched by computerization; the leading role in this development will continue to be played by the creative engineer.

Since the early 1970s, the CAD/CAM concept has been used to describe complex integrated solutions of computer-aided processing from design to production planning and to increasingly automated manufacturing. Although the industrial application of such complex solutions has attained a high level, there are still specific tasks awaiting solution. These are mostly connected with the definition of interfaces, the structure of data bases and largely standardized data handling.

International efforts aimed at defining standardized product information (see chapter III) are of great importance. Based on proposals for standardization, such as IGES (United States), SET (France) and PDES (ISO), the projects under development are significant for the implementation of:

- Data transfer between several CAD and CAM departments within the factory;
- National data transfer between companies; and
- International co-operation and transfer of technology.

Special attention should therefore be given to work in international standardization foreseen by ISO up to 1990. The solution finally adopted should particularly address the requirements of small and medium firms, whose resources are often limited.

The efforts being made in all industrialized countries towards computer-integrated plants (CIM) are of great significance at present. In advertising, mention is made of CIM solutions. Realistic international forecasts maintain that the first plants with CIM will be in practical operation by between 1992 and 1995. However, independent of these forecasts, CIM principles should already be receiving special attention.

CIM represents the currently highest level of integration and informational interlinking of all working departments in a factory (see chapter III): a network connecting all the computers and automated equipment. The accomplishment of this complex task demands comprehensive efforts in the field of international standardization. A first approach and proposal in this direction was presented with the manufacturing automation protocol (MAP) meeting the requirements of the ISO reference model.

In connection with the development towards CIM structures, the integration of several kinds of knowledge support systems is significant. International activities in artificial intelligence have not as yet resulted in practical applications. Assuming further development of the theoretical framework and of electronic data processing, the integration of new kinds of knowledge processing may be expected. This should include the following aspects:

- Data processing,
- Information processing, and, finally,
- Knowledge processing.

Knowledge processing in the technical branches is characterized by the use of expert systems and by machine intelligence.

Software engineering productivity and software reliability can be considerably improved by using appropriate software tools. Advanced software development systems supporting programming, testing, debugging and program documentation phases of software production become increasingly available.

The effects of computer-aided modes of activity are determined by the level of qualification of the staff. The rapidly increasing availability of hardware has resulted in a lack of staff qualified in its proper manipulation. These lacunae cannot be filled in the short run by the universities and colleges alone. Factories must organize their own capacities for further training. In addition to the development of software solutions, there is need for an integrated qualification strategy. In that context, the computer should be used not only as a subject of teaching, but also as a means of training.

The most important effect of CAD/CAM is its strong influence on the competitiveness of products. CAD/CAM systems may also be evaluated by means of multicriteria comparison. This solution is, however, not yet sufficiently elaborated because certain qualitative variables cannot be quantified. The main advantages of CAD/CAM systems are:

- Reduction of lead time and of time needed to manufacture one part;

- Assurance of the terms of delivery in higher quality;

- Improved scheduling of manufacturing equipment and reduction of stocks;

- Increased transparency through improved organizational co-ordination of sequences; and

- Possibility of carrying out simulation and calculation of variants at the stage of production planning.

Illegal software transfer is a serious problem for software companies and causes them considerable losses. The need is generally recognized for software protection, both nationally and internationally. International consensus is emerging that copyright protection is needed to secure software products. However, the views of legal experts and government authorities still diverge over the best way to achieve this. In this respect, the work of the WIPO/UNESCO group of experts on copyright aspects of the protection of computer software is of particular interest.

In all industrialized countries, governments have taken measures to promote the development and introduction of industrial automation with particular emphasis on the software aspect. Comprehensive national programmes have been designed comprising direct financial assistance, R and D grants and subsidies and indirect benefits in tax policies. Funding arrangements concentrating on both technologies and specific industry sectors are dedicated to users and producers.

Intergovernmental co-operation is focused on the establishment of an adequate framework for R and D and production of industrial automation technologies. Efforts should be made to facilitate international co-operation of firms, universities, research institutes, etc. The role of the numerous non-governmental organizations active in the field is to promote and co-ordinate the exchange of information. The work of the organizations dealing with standardization (ISO, IEC) is of particular importance as they provide the necessary conditions for improved international co-operation, specialization and development.

ANNEX I. DEFINITIONS OF RELEVANCE TO THE PRESENT STUDY

A. SELECTION OF KEY DATA-PROCESSING TERMS

Applications software

- Programs and packages designed to satisfy applications. Contrasted with systems software [13].

- All programs whose purpose is to solve the computer user's own problems [3].

Applications software includes the programs created for the computer-aided solving of technical, economic and organizational problems or for controlling automated processes. The present study, in agreement with the basic objective "software for industrial automation" focuses on applications software.

Artificial intelligence (AI)

- The ability of any machine or routine to learn and improve its performance as a result of the repetitive execution of a given activity or search for solutions to a given set of problems [13].

- The study of computer techniques to supplement the intellectual capabilities of humans. All is concerned with the more effective use of digital computers through improved programming methods [19].

Debugging

- To locate and correct any syntactic and semantic errors in a computer program; and

- To detect and correct malfunctions in the computer itself (related to diagnostic routine).

Firmware

- (i) A program blown into ROM; logic which has been permanently hard-wired; distinguished from hardware only by the fact that the functions performed were once carried out by software.
  (ii) Software which interacts intimately with hardware carrying out essential systems software functions [13].

- One important technological variant in systems software which is incorporated in the hardware. This is usually done in "read only memories" forming part of the physical structure of a computer, where the software concerned is recorded once and for all [3].

- 149 -

Hardware

-   Physical equipment, as opposed to programs, procedures, rules and associated documentation. (International Organization for Standardization [12]).

Hardware is the general term for all kinds of information-processing equipment which, in connection with industrial automation, is employed in various forms of use, i.e.:

-   Control units for automated machine tools, equipment and devices;

-   Process computers for the controlling of complex technological processes, the storage and time-adequate distribution of control data and others;

-   Engineering work-stations, i.e. micro- and minicomputers as independent computers with a special peripheral equipment (input and output devices, graphic and alphanumeric display screens, plotters and others); and

-   Electronic data-processing (EDP) systems of various sizes and capacities.

Integrated software systems

-   Integrated software systems represent the most important technological trend in software for the 1980s. They are so designed that the different systems and application software components match one another, sharing data and transferring results among the various programs on just a particular site or on a number of sites communicating with one another.

-   The tendency towards integration gives rise to four types of systems: the integrated system on a central mainframe, the integrated system on a microcomputer, the distributed system involving a central mainframe and microcomputers, and office automation systems. A fifth type of software, communication software - and particularly local area network software - plays an infrastructural role in promoting the last two applications [3].

Interface

-   A shared boundary between two functional units, defined by functional characteristics, common physical interconnection characteristics, signal characteristics, and other characteristics, as appropriate [15].

-   (i) A hardware and/or software link which enables two systems, or a system and its peripherals, to operate as a single, integrated system.

-   (ii) The input devices and visual feedback capabilities which allow
         bilateral communication between the user and the system. The
         interface to a large computer can be a communication link
         (hardware), or a combination of software and hard-wired
         connections. An interface might be a portion of storage
         accessed by two or more programs, or a link between two
         subroutines in a program [16].

Local area networks (LAN)

-   A LAN serves internal communication. It connects workplaces and
    workplace systems with central systems and external networks.
    Communication media employed are copper wires, coaxial cable and
    optical fibres; types of transmission employed are broadband and
    baseband. LAN are typically limited to private territory:
    distances are less than 10 km, usually less than 2 km. A LAN may be
    a subnetwork, connected through gateways to other public or private
    networks, thus forming a global network [14].

-   A LAN is a system of hardware and software that allows members of
    logically related groups to communicate over distances of a few feet
    to 20 miles (32 kms) or more. In the factory, typical group members
    may include robots, programmable controllers, data collection
    devices, video monitoring systems, and badge readers. By
    facilitating machine-to-machine communication, LANs can weld the
    different phases of a manufacturing operation into a single, unified
    process [22].

Open system

A system whose characteristics comply with specified standards and that
therefore can be readily connected to other systems that comply with the
same standards [17].

Open Systems Interconnection - Basic Reference Model

In 1984, the International Organization for Standardization (ISO) adopted
the Open Systems Interconnection (OSI) Basic Reference Model
(International Standard ISO 7498). This International Standard
establishes a framework for co-ordinating the development of existing and
future standards for the interconnection of systems and is provided for
reference by those standards.

The general structure of the OSI architecture as described in the above
International Standard provides architectural concepts from which the
Reference Model of OSI has been derived.

- 151 -

The Reference Model contains seven layers

  (i)   Application Layer (layer 7);
  (ii)   Presentation Layer (layer 6);
  (iii)   Session Layer (layer 5);
  (iv)   Transport Layer (layer 4);
  (v)   Network Layer (layer 3);
  (vi)   Data Link Layer (layer 2); and
  (vii)   Physical Layer (layer 1);

These layers are described in more detail in chapter II.3. The highest is the
Application Layer and it consists of the application-entities that co-operate
in the OSI environment. The lower layers provide the services through which
the application-entities co-operate [11].

Software

  -   International Organization for Standardization (ISO):

      (i)   Intellectual creation comprising the programs, procedures,
             rules and any associated documentation pertaining to the
             operation of a data-processing system.

      (ii)   Software is independent of the carrier used for transport
             [12].

  -   World Intellectual Property Organization (WIPO):

      (i)   "Computer program" means a set of instructions capable, when
             incorporated in a machine-readable medium, of causing a
             machine having information-processing capabilities to
             indicate, perform or achieve a particular function, task or
             result.

      (ii)   "Program description" means a complete procedural
             presentation in verbal, schematic or other form, in
             sufficient detail to determine a set of instructions
             constituting a corresponding computer program.

      (iii) "Supporting material" means any material, other than a
             computer program or a program description, created for aiding
             the understanding or application of a computer program, for
             example problem descriptions and user instructions.

      (iv)   "Computer software" means any or several of the items
             referred to in (i) to (iii) [24].

Systems software

  -   Intimate software, including operating systems, compilers and
      utility software. Contrasted with applications software [13].

-       The combination of programs required to make optimum use of the
        computer and its peripherals. In the first case it includes
        operating systems which handle the resources of a computer system.
        Next come compilers and interpreters which translate programs
        written and stored in the different symbolic languages (COBOL,
        FORTRAN, Pascal, etc.) into sets of operational instructions that
        can be interpreted and performed by the computer. Systems software
        also includes data-base management systems and utility programs and
        debugging aids [3].

-       Systems software includes all the programs and program components
        which are necessary for the realization of basic functions and the
        user-adequate application of computers and means of automation.
        These comprise, especially:

        (a)   Operating systems, UNIX being a generally applicable
                operating system;

        (b)   Compilers for higher programming languages, such as FORTRAN,
                ALGOL, PL/1, BASIC, MODULA, LISP, PROLOG and others, FORTRAN
                77 being most frequently used in engineering;

        (c)   Graphics software; and

        (d)   Library programs, especially for mathematical functions.

Testing

-       The verification that the computer program actually realizes the
        control algorithm. Testing starts with the definition of testing
        objectives and ends with a well documented record of test results,
        so that re-testing following modifications can be minimized.

Tuning

-       The adjustment of control constants in algorithms or analogue
        controllers to produce the desired control effect.

B.   SELECTION OF KEY INDUSTRIAL AUTOMATION TERMS

Adaptive control system (ACS)

A control system which continuously monitors its own behaviour and, by adjusting its parameters, is able to adapt itself to a changing environment.

Automated guided vehicles (AGV)

Vehicles equipped with automatic guidance equipment which follow a prescribed path which interfaces with workstations for automatic or manual loading and unloading [16].

Automated material-handling system

Systems used to automatically move and store parts and raw materials throughout the manufacturing process and to integrate the flow of work pieces and tools into the manufacturing process [16].

Automated storage and retrieval system (AS/RS)

A high-density rack storage system with rail running vehicles serving the rack structure for automatic loading and unloading. Vehicles interface with AGV systems, car-on-track, towline or other conveyor systems for automatic storage and retrieval of loads [16].

Basic software for CAD/CAM solutions

Basic software for CAD/CAM solutions is a specific part of systems software designed to improve its adaptability, portability, transparency and user friendliness. Basic software for CAD/CAM work-stations assists the user in his interactive work, supporting data-handling, data-base-management operations, dialogue functions, etc. [85].

Computer-aided design (CAD)

-   The application of computers to design where the designer converses directly with the computer by using a graphic or nongraphic console in such a manner that his problem-solving processes are highly responsive and essentially uninterrupted [7].

-   A system which incorporates one or more computers for carrying out some of the calculations and actions involved in the design process (CECIMO Working Party on Standardization) [2].

Computer-aided manufacturing (CAM)

-   The application of computers to various manufacturing operations including material and information flow.

-   A system which incorporates one or more computers for carrying out some of the tasks involved in the organization, scheduling and control of the operations involved in the manufacture of the product. Where machining is involved, a CAM system will usually involve CNC-machine tools and means for producing part programs for them and it may also involve a central computer for scheduling, planning and control of the operation of the system. It may involve a DNC-system using either the central computer or a separate computer, as well as computer control of stores, orders etc. (CECIMO Working Party on Standardisation) [2].

Computer-integrated manufacturing (CIM)

-   A closed-loop feedback system in which the prime inputs are product requirements (needs) and product concepts (creativity) and the prime outputs are finished products (fully assembled, inspected and ready for use). It comprises a combination of software and hardware, the elements of which include product design (for production), production planning (programming), production control (feedback, supervisory and adaptive optimizing), production equipment (including machine tools) and production processes (removal, forming, and consolidation) [20].

-   The concept of a totally automated factory in which all manufacturing processes are integrated and controlled by a CAD/CAM system. CIM enables production planners and schedulers, shop-floor foremen and accountants to use the same data base as product designers and engineers [16].

Computerized numerical control (CNC)

-   A numerical control system wherein a dedicated, stored program is used to perform some or all of the basic numerial-control functions [18].

-   A technique in which a machine-tool control uses a computer to store NC instructions generated earler by CAD/CAM for controlling the machine [16].

Direct numerical control (DNC)

- A system connecting a set of numerically controlled machines to a common memory for part program or machine program storage with provision for on-demand distribution of data to the machines [18].

- A system in which sets of NC-machines are connected to a mainframe computer to establish a direct interface between the DNC computer memory and the machine tools. The machine tools are directly controlled by the computer without the use of tape [16].

- Two kinds of DNC have been developed: DNC-MTC (machine-tool controller) and DNC-BTR (behind the tape reader):

  DNC-MTC: Developed in the 1960s and early 1970s (the original DNC). The central computer controls directly the individual machines.

  DNC-BTR: Each machine tool has it own control unit but receives its program instructions from the central computer, which is the program library for the machine system and supervises the individual machine operations by "go and no go" instructions [2].

Flexible manufacturing unit (FMU)

A one-machine system, usually a machining centre or a turning centre, equipped with a multi-pallet magazine, an automatic pallet changer or robot and an automatic tool-changing device. The unit is able to operate partly unattended [2].

Flexible manufacturing cell (FMC)

A system, which comprises two or more machines usually at least one machining centre or turning centre, multi-pallet magazines and automatic pallet and tool changers for each machine. All machines, as well as the operations carried out by the cell, are controlled by a DNC-computer [2].

Flexible manufacturing system (FMS)

An integrated computer-controlled complex of numerically controlled machine tools, automated material and tool-handling devices and automated measuring and testing equipment that, with a minimum of manual intervention and short change-over time, can process any product belonging to certain specified families of products within its stated capability and to a predetermined scheduled. A FMS is made up of two or more FMC connected by an automatic transportation system (automated guided vehicles, computer-controlled cranes, etc.) which moves pallets, workpieces and tools between machines and to and from workpiece and tool storage. The whole system is under the control of a DNC-computer which is usually connected to a factory host computer [2].

Industrial robot (IR)

Automatic position-controlled reprogrammable, multi-functional manipulator having several degrees of freedom capable of handling materials, parts, tools, or specialized devices through variable programmed motions for the performance of a variety of tasks [1].

Machining centre (MC)

- A machine capable of performing a variety of metal-removal operations on a part, usually under numerical control [16].

- A machine conceived for flexibility, which can work on a variety of pieces, using multiple-axis control to accommodate complex parts and calling on potentially hundreds of tools to rough and finish face, bore, drill, tap, and the like [85].

Material requirements planning (MRP)

A full-fledged manufacturing information system which, by means of a closed-loop feedback from the shop floor, provides continuous information on production requirements and on the manufacturing process. The inputs into an MRP system are a master production schedule, bill of material, and inventory balances. The output is a production schedule [7].

Numerical control (NC)

- Automatic control of a process performed by a device that makes use of numeric data usually introduced while the operation is in progress. The term numerical control is commonly used in machine-tool applications [12].

-       A numerical control system is a hierarchically arranged quantity of
        signal-interpreting, signal-storing, signal-processing, and
        signal-transferring elements, where the signals are offered in
        different forms: as letters, to describe geometric and
        technological circumstances and to define axes of machines on
        orders; as dimension figures, to describe the geometry; as
        characteristic values, to input coded information [23].

Programmable controller:

A solid-state control system that can be programmed to execute
instructions that control machines and the process operation by
implementing specific functions such as logic control, sequencing,
timing, counting and arithmetic operations. A programmable
controller consists of five basic elements: (1)central processing
unit, (2)memory, (3)input/output interface, (4) power supply, and
(5) programming device [16].

ANNEX II.   PLANNING AND RESCHEDULING OF FMS 1/

Scheduling, which ranks amongst basic problems in the area of FMS control, has been undertaken by computers since the beginning of their application in manufacturing control. The contemporary state of the use of computers in planning and rescheduling depends, above all, on the possibility of interconnecting planning systems with other control subsystems by means of a common and uniform data base (e.g. data bank). Feedback from the monitoring of the manufacturing process is also important, as is the possibility of integrating FMS control systems and whole management information systems into integrated systems, e.g. within the framework of CIM.

Employment of a common data base makes it possible to analyse or assess manufacturing processes on a continuous basis and to link automated functions - for instance material-requirement planning. Typical program packages, whose functional framework also covers, to a certain extent, the problems of planning and rescheduling of current FMS, are for instance PICS from IBM, BASIS from Siemens (Federal Republic of Germany), PLUS from Robotron (German Democratic Republic) or MARS, resp. VARS (first stage) from Kancelarské stroje (Czechoslovakia). A qualitative change has occurred in computer applications in the area of planning, rescheduling and control of FMS with the introduction of computers equipped for user dialogue and providing real-time operation including an up-to-date picture of the actual state of the manufacturing process. Process scheduling can be brought into real time, with input from individual manufacturing cells. In addition, this makes it possible to interconnect the scheduling systems with direct control systems, including the automated handling of materials and finished products.

The application of interactive data processing, the increased capacity of computers and real-time processing also permit quality improvement in planning and management, both at the level of individual FMS and at that of the whole enterprise, owing to the employment of simulations for the evaluation of partial decision impacts - e.g. acceptance of additional orders - on the whole manufacturing process. These program systems include COPICS from IBM, or (with certain limitations) VARS (second stage) from Kancelarské stroje.

4.   Beside compact systems for planning and control, including connected subsystems, numerous program packages are available which ensure or improve planning tasks in FMS. Examples of these include CAPOS from IBM or the rescheduling system Detail Schedule - Dynamic Schedule, developed by INORGA (Czechoslovakia).

The latest developments in this area suggest that planning and rescheduling of FMS will in future be even more integrated into the whole system of management, eventually in the framework of CIM. In this way, planning and rescheduling will cover not only the area of main production, but also the entire connected material and information flow of a given manufacturing complex. The employment of multifunctional reprogrammable robots, CNC or DNC machines will significantly simplify the process of planning and management. As regards single-product and small-batch production, the application of effective planning systems ensures that only the necessary parts will be produced and that stocking will be minimized.

1/ Material transmitted by the Government of Czechoslovakia.

ANNEX III: SIMULATION PROGRAMMING SYSTEMS FOR FMS 1/

Introduction

The use of simulations for studying FMS is of considerable importance. Reliable simulation models can help users both in making forecasts and in the control of manufacturing processes by providing information to facilitate objective decision-making, investment allocation, verification of production plans, product mix, etc. The basic version of the simulation model described below is expected to be programmed, debugged and tested on at least one operating FMS by the end of 1986.

Design of a simulation model

From the point of view of computer modelling, FMS for non-rotary parts have many common characteristics. As regards the system operation and control, there are two significant material flows, the flow of pallets and the flow of tools.

The system which has been studied incorporates two types of elements. The first type covers permanent activities (A) characterized by the fact that they are present in the system for the whole time of the simulation. In the case of FMS, such features (A) include manufacturing cells and conveyers. The second type are the so-called transactions (transition features (T)), which exist for a certain time only. The number of those features is variable. In the case under consideration the transactions (T) include, for example, pallets, workpieces, tools, etc.

In general, it could be said that there may exist so-called A-systems (composed only of permanent features), T-systems (composed only of transition features), AT-systems (composed of both, and where action is carried by the transition feature) and, by analogy, TA-systems, where the action is carried by permanent features. The above-mentioned different "views" on the system have particular significance for modelling and for the selection of a suitable simulation language.

How to model FMS

An analysis of the production technology for non-rotary parts and the assortment produced in those manufacturing sections leads to the conclusion that the most suitable solution is to model FMS as a TA-system. That means that the "overall action" of the modelled process has to be broken down into the various actions, describing activities of individual manufacturing cells.

The pallets and workpieces are moved through the system by shifting the pallets from one set to the next. The pallets are shifted as a passive system.

1/ Material transmitted by the Government of Czechoslovakia.

For the majority of manufacturing cells the following operating cycle is valid:

- Wait until the work storage is filled;

- Block the proper (usually the first) pallet with workpiece;

- Wait for the conveyer;

- Pass the pallet to the machine (either from store or from the preceding manufacturing cell);

- Execute proper operations according to the technological process;

- Wait for the conveyer;

- Pass the pallet into the "queue" to the next manufacturing cell (if the following manufacturing cell is free, then directly to it, otherwise to the store);

- Transmit the signal that the pallet has been sent to the proper group of interchangeable manufacturing cells;

- Repeat the cycle starting from the first item.

It should be remembered that the modelled system is highly interrogative. It is also useful if several conditions can be simulated simultaneously.

It is necessary to use algorithms of simulation languages for the selection and scheduling of each event and time shift. This makes it possible to execute a simulation in an "event to event" mode, i.e. to skip over "idle intervals".

All the above-mentioned requirements are met by the universal programming language SIMULA 67, which provides means for discrete simulation and possibilities of various sights of the system (A, T, AT, TA and others). The language can work with files, compute the initial amount of pallets and allocate tools to tools-unit storage. A very important aspect is the possibility of using classes (which has been developed in SIMULA) for heuristic optimization of the simulation models and for the simulation models themselves.

Suggested simulation method

The starting point for structuring the algorithm is the determination of external and internal parameters.

Parameters of FMS assumed as external (Technological constants, which characterize production ensured by FMS):

- Number of similar types of parts produced;

- Duration of operations;

- Number of clampings of parts necessary for complete machining;

- Number of different tools (i.e. various kinds and dimensions) necessary for the production;

- Machine constants of the system;

- Numbers of types of working stations, i.e. number of groups of interchangeable manufacturing cells in the system according to the technological process;

- Resetting of pallets and tool-transport system constants.

Parameters of FMS assumed as internal

- Number of interchangeable manufacturing cells inside the groups;

- Number of identical specializations of technological pallets for identical clampings;

- Number of identical tools for operation of the system; initial tool equipment of FMS;

- Shift configuration of manufacturing cells.

The external parameters are entirely independent of strategies and material-flow control concepts. The internal parameters are optional within the framework of reassuming simulation experiments and create starting conditions for the "tuning" of the system, which essentially influences the results.

Within the framework of experimenting - e.g. in the case of systems design - some external parameters (store capacities, pallets capacities, speed of conveyers, etc.) are also assumed as the criterial function components (beside the criterial function components already established, such as maximum capacity of the manufacturing cells, maximum production, etc.) in order to obtain values as indexes for designers. At the same time, it is assumed that dynamic simulation with the external and internal parameters as well as the capacity parameters proceed simultaneously, as provided for in the SIMULA classes PAROPTIM and PAROPTMULTI.

The choice of criterion, if the user is able to formulate the criterion in the form of an algorithm, can also be accomplished during the solution (or use) of the simulation model. To clarify extreme states, it is necessary to follow up the behaviour of the system during the manufacturing programs, which makes special demands on the system. During the verification of the system's permeability for a selected programme, it is possible to tune the system (by means of internal parameters), as well as to specify limiting values for selected characteristics (pallet store capacity, speed of conveyers, etc.) by means of multicriterial optimization.

ANNEX IV: DESCRIPTION OF AN INDUSTRIAL MANUFACTURING CELL CONTROLLER 1/

Introduction

The term manufacturing cell here denotes a group of one or more machine tools (e.g. lathes, milling machines, machining-centres and electric discharge, grinding, boring and drilling machines), adjacent mechanical elements (such as work-piece buffers or a chip-disposal system), tool- and fixture-supporting work-stations and an optional measuring machine. This group of machines can operate as a stand-alone unit under the control of a local computer. The cell has a limited ability of failure-detection and a strategy for recovery. A manufacturing system may contain one or more cells of this kind.

The cell controller is based on an industrial 16-bit micro-computer. Its high-level program-support makes it a reliable device for factory automation.

In a standard manufacturing cell, the following elements can be connected to the cell controller:

- Technological machine (with controller);

- Workpiece and tool-storage equipment (with controllers);

- Cleaning and burring stations (with controllers);

- Transport systems for in-cell and inter-cell material transfers; and

- In-cell inspection (e.g. measuring machine) station.

The main functions of the cell controller are the following:

- Task scheduling;

- Operation according to the schedule;

- Failure detection and diagnosis; and

- Process-data monitoring.

The capability of the system depends on the actual configuration.

1/ Extracted from a publication of the Computer and Automation Institute of the Hungarian Academy of Sciences in Budapest.

Task scheduling

Task scheduling covers the generation of a task and timing table for all cell elements on the basis of information supplied by the shop-floor production planning. In this way, optimum machine-load or optimum processing-time may be calculated. In case of the breakdown of an element in the cell, the scheduler redistributes the tasks amongst the operating elements. The cell controller is able to rearrange the configuration without human interaction if the necessary information has previously been provided. (Configuration of the cell may be necessary on several occasions. The cell control software provides fast and flexible utilities for this task).

Operation according to the schedule

This function entails the co-ordination of various activities of the cell elements (in the case of normal operation), the management of parts, tools, pallets and jigs, and the provision to the elements of the necessary information.

Beside the co-ordination of operations, the cell controller enables the supervisor (or operator) to obtain various technological and process data or to modify some stored parameters. These data are also available for a higher control level (central control computer).

Failure detection and diagnostics

When an abnormal event is reported to the cell controller (e.g. an accident or the failure of a component), diagnostic functions are activated. After detecting the error, the following activities can be initiated:

-   Correction steps to avoid additional damage;

-   A series of diagnostic tests to locate the cause of the error; and

-   Selection of alternative technologies, if possible, to continue the operation.

Process-data monitoring

The important events during operation are automatically logged but are not directly available to the user. This logging is the basis of the user logging, where the user has the possibility of selecting certain events, which will be recorded in a separate log file.

Special features of the cell controller

Software

The software of the cell controller is based on an advanced operating system. The functions are performed by structured modules developed in a high-level programming language. The architecture of the software makes possible the application of standard data-base management systems.

Hardware

The cell controller has direct links to every cell element, with the information passing only through these communication lines. The controllers of cell elements should have a certain set of DNC functions.

The cell and its environment

The flexible manufacturing cell makes possible the integration of computer-aided design and computer-aided manufacturing CAD/CAM into a comprehensive system. The output of a CAD system and a computer-aided process-planning system can be connected directly to the flexible cell controller, and thus the elimination of tedious human work becomes possible.

ANNEX V: A STANDARD ARCHITECTURE OF SOFTWARE FOR FMS 1/

Introduction

The most important element of any FMS is its computer-aided control system (CAC FMS), which ensures the rational organization of the main and subsidiary manufacturing operations, automatically records and regulates the manufacturing process, and rapidly adjusts that process to the production of new types of goods without the need to replace the manufacturing equipment.

Quite a lot of experience has now been acquired of developing, debugging and testing programs for industrial computer-aided control systems. However, the development of CAC FMS has a number of special features that weigh heavily on software design. A CAC FMS operates principally in real time, which means that provision must be made not only for rapid processing of the system data but also for its updating with no interruption of the manufacturing process. The hardware for a CAC FMS comprises a co-ordinated set of mini- and micro-computers of varying configurations and, perhaps, with varying base and system software. Consequently, the local computers that control individual components of the FMS must be made to interact with the main computer that controls the system as a whole, and account must be taken in this respect of the differences in the computing process on these various types of machine. Because the operation of an FMS must be highly reliable, the software for a CAC FMS must include provision for the redistribution of control functions and part-processing modes (reconfiguring the system) in the event of the maintenance or failure of computing or manufacturing equipment. The software for an FMS therefore constitutes a multi-purpose hierarchical system that is characterized by the difficulty of linking and integrating its individual program modules into a whole. Consequently, given the stage now reached in the development of FMS, it is obviously urgent to standardize the software for FMS in so far as the structure of the CAC, the distribution of control functions amongst the various levels, and the functional characteristics of each of the elements of the system will permit.

Success in standardizing software would both cut system design costs and time and yield better-designed and more reliable CAC FMS.

A Standard architecture of a computer-aided control system for an FMS

A computer-aided control system for an FMS typically comprises three interacting hierarchical levels (figure V.1):

- Scheduling;

- Supervision; and

- Direct control of process equipment.

1/ Material transmitted by the Government of the USSR.

Figure V.1. Standard architecture of CAC FMS

Scheduling
1. Support of information interfaces with CAECS, CAPES AND CADCP.
2. Support of CAC FMS data base.
3. Planning of production down to level of shift targets.
4. Recording of fulfilment of shift targets.
5. Compilation and support of control-program libraries.

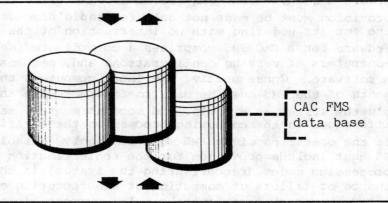

CAC FMS
data base

Supervision
1. Co-ordination and monitoring of work of FMS equipment.
2. Distribution of control programs.
3. Collection and recording of data on progress of work.
4. Supervision of progress of work.
5. Reconfiguration of system in event of equipment failure.

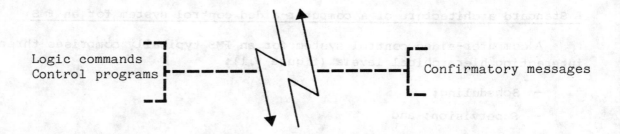

Logic commands
Control programs

Confirmatory messages

Local control of process equipment
1. Reception and decoding of control programs.
2. Formulation and transmission of commands to activating systems.
3. Reception of messages from activating systems.
4. Monitoring of functioning of FMS equipment.
5. Communication with FMS staff.

The interaction between the scheduling and supervision levels takes place as follows: every day (every shift), the supervision level receives from the scheduling level production assignments arranged according to the priority of their completion and the availability of resources within the system and sends back to the scheduling level information on which assignments have been completed and which not and which resources remain unused.

A CAC FMS generally forms part of an integrated enterprise-control system (computer-integrated manufacturing system), the components of which provide the CAC FMS with the essential information. Then the scheduling level in the CAC FMS must also interact with the enterprise-control level, the computer-aided production-engineering system (CAPES) and the system for the computer-aided design of control programs for numerically-controlled equipment (CADCP). The computer-aided enterprise-control system (CAECS) transmits to the CAC FMS plans for the production of goods by the FMS, data about the composition and engineering and operating characteristics of the equipment in the FMS and scheduling standards. The CAPES transmits to the FMS data on process paths and standards for the various manufacturing operations. The CADCP provides control programs for the numerically controlled manufacturing and measuring equipment.

The scheduling level of the CAC FMS interacts with the level of direct control of the process equipment - which encompasses the control of product manufacture, the supply to and removal from the manufacturing zone of parts and tools, and the control of transport, storage and measuring devices - and with terminals at production engineers' and FMS staff's work-stations. In particular, the level of direct control of process equipment receives instructions for the dispatch from storage, the transport and the processing of individual workpieces. The supervision level receives messages concerning the results of the execution of its instructions.

Associated with each of the levels in the CAC FMS hierarchy are particular control functions. The most important of those functions are described below.

The basic functions of a CAC FMS

The operations effected at the scheduling level of a CAC FMS are the following:

- Building and maintenance of the system's data base;
- Elaboration of the monthly timetable for the manufacturing of batches of products;
- Elaboration of the sliding batch-work schedule and of work-preparation plans for the specified number of days (10 or 5); and
- Calculation of the targets per shift for each type of manufacturing operation and recording of the level of compliance with those targets.

At this level, plans and targets are determined on the basis of technical and economic calculations and the results of optimization exercises. Functions relating to the recording of the progress of work are effected on the basis of the primary records compiled at the supervision level.

The scheduling level is also the level for the creation of the long-term and short-term (immediate use) libraries of control programs.

The functions of the supervision level have to do with the organization of the automatic passage of orders through the FMS in keeping with the targets set for each operation and each shift. At this level, the following functions must be performed in real time:

- Control of the operation of the machining centres and of the automatic transport and storage systems, including the co-ordination of their work and preservation of their interaction as the manufacturing process progresses;

- The collection, processing and presentation of information on the fulfilment by the FMS of the targets set for each shift, including the compilation of records of finished products and rejects and of the presence and movement of workpieces, tools and fixtures in stores and at work-stations;

- The collection, processing and presentation of information on the condition of process equipment, tools and fixtures, including the keeping of records of idle time for equipment, tools and fixtures, records of equipment downtime by cause and records of tool and fixture service life, and the diagnosis of the fitness of equipment for operation; and

- Interfacing between the FMS's human supervisor and the CAC FMS, including receiving and processing inquiries from the supervisor, transmitting to him various messages on the progress of work, and serving as a channel for intervention by the supervisor in the manufacturing process.

The function of direct control of the manufacturing equipment includes the reception and decoding of control programs, the formation of control commands for local machine-tool control systems and automated transport and storage systems, the collection of information on the operation of such systems, and the transmission of control information to the supervisory staff. It should be noted that the number of functions executed at this level of control depends essentially on the manufacturing equipment used in the system. In the more complex systems, the functions at the level of direct control may include storage of control programs in shop-floor computers, updating of such programs, automatic monitoring of equipment and its local control systems, and optimization of rates of work or routes of transport equipment, etc. Another distinguishing feature of this level of control is the sharing of control functions between the central computer of the CAC FMS and the local manufacturing-equipment control systems. The precise extent to which the functions are shared depends on the software for the local control systems.

- 172 -

The standard architecture of software for CAC FMS

In keeping with the organizational structure of a CAC FMS described above, control of an FMS is effected by means of a multipurpose three-level organization and control system. Based on an analysis of experience in the development, introduction and operation of such systems, the main principles for the determination of the structure and content of standard software for CAC FMS are stated below and the requirements specified as regards the capabilities of its various components.

For any CAC FMS, the software comprises base, system and functional software (figure V.2).

The base software of a CAC FMS includes the operating systems, programming languages, translators, service programs and the drives of standard devices. It should be noted that the capabilities of the base software depend to a large degree on the capabilities of the software supplied with the computers used in the FMS.

The system software includes the service programs for the computing networks, the means of organizing and maintaining the CAC FMS data base, and optimization and simulation methods.

The computer-network service programs are designed to unite the differing computing equipment used at the various levels of control of the FMS into a single system.

Whatever its end product may be, the efficiency of an FMS is highly dependent on the extent to which the work schedule has been optimized in the light of the order of priority attached to the various products and the interchangeability of the manufacturing equipment. Schedule optimization is the function of the special programs for the execution of optimization techniques.

The core of a CAC FMS is its data base. This contains data on the range of products and the time required to make them, on the availability of resources (equipment, tools, fixtures, blanks, semi-manufactures, accessories, etc.) and their condition at all stages of the manufacturing cycle, and data on the main and alternative equipment, etc. The reliability of operation of a CAC FMS data base largely determines the reliability of operation of the system as a a whole.

The main functions of CAC FMS data-base organization and support software are to load the base, update it in the batch and interactive modes, ensure interaction between the CAC FMS data base and the data bases for computer-aided enterprise control, CADCP and CAPES, and to process requests for data from the functional software of the CAC FMS and the FMS supervisory personnel.

The software for organizing and maintaining a CAC FMS data base is divided into two functionally distinct sections. The first, the design section, includes the algorithms and programs which, on the basis of the parameters describing a specific FMS, execute machine design of the variable program modules for the expected use conditions. The second, the executive section, contains interpretive programs that, with the help of the variable program modules, execute the functions of organizing and maintaining the data base in question.

Figure V.2. Standard architecture of CAC FMS software

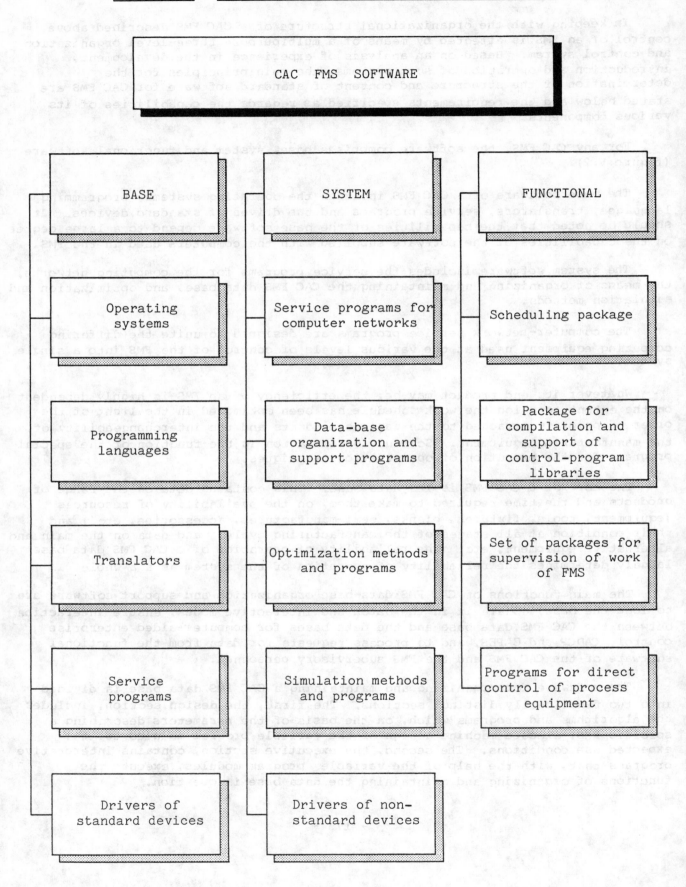

CAC   FMS   SOFTWARE

BASE	SYSTEM	FUNCTIONAL
Operating systems	Service programs for computer networks	Scheduling package
Programming languages	Data-base organization and support programs	Package for compilation and support of control-program libraries
Translators	Optimization methods and programs	Set of packages for supervision of work of FMS
Service programs	Simulation methods and programs	Programs for direct control of process equipment
Drivers of standard devices	Drivers of non-standard devices	

The need to include simulation facilities in the software of a CAC FMS derives above all from the high capital-intensiveness of FMS and the need to resolve the difficult problems of analysing, designing and optimizing the control system for a FMS before the system itself is set up. Simulation is the most effective means of achieving such goals.

By using simulation techniques during the design of an FMS, it is possible to determine with sufficient accuracy for practical purposes the main qualitative parameters and the architecture of the system. This applies, for example, with respect to the assessment of the productivity of the manufacturing equipment and of the duration of the complete manufacturing cycle, to the selection of the most appropriate combination of process machinery, to the determination of the necessary storage capacity, etc.

As regards the debugging and testing of a CAC FMS, simulation makes it possible to undertake separate or joint debugging of individual control subsystems, and to test and demonstrate the capabilities of the entire control system without any need for the manufacturing equipment to be available.

When it comes to the operation and modernization of a CAC FMS, models of the kind in question can be used to develop algorithms and programs for initiating new events in the production process and, when necessary, for adapting the control system to changes in the technological parameters of the FMS with a minimum of delay.

The functional software in a CAC FMS includes a set of application packages and program modules executing the functions described above and grouped according to the standard organizational structure of a CAC FMS (figure V.3). The "monthly planning of FMS output" package fixes the order for the initiation of work on batches of products over the course of the month and works out the monthly schedule for each type of manufacturing operation.

The calculation and updating of the operation schedules and of the work-preparation plans for specific numbers of days, as well as the preparation of plans for the supply of blanks, tools and fixtures, are performed by the "planning of FMS operation for specified period" packages.

The "per-shift and per-day planning of FMS operation" package automatically calculates the targets for each type of manufacturing operation per shift and draws up a plan for the supply of blanks, semi-manufactures, accessories, tools and fixtures for each 24-hour period.

The "organization and support of control-program libraries" package works on the basis of the monthly schedule and the per-shift targets to prepare long-term and short-term libraries of control programs. The long-term libraries are created by selecting the necessary programs from a CADCP data base.

Figure V.3. Architecture of CAC FMS functional software

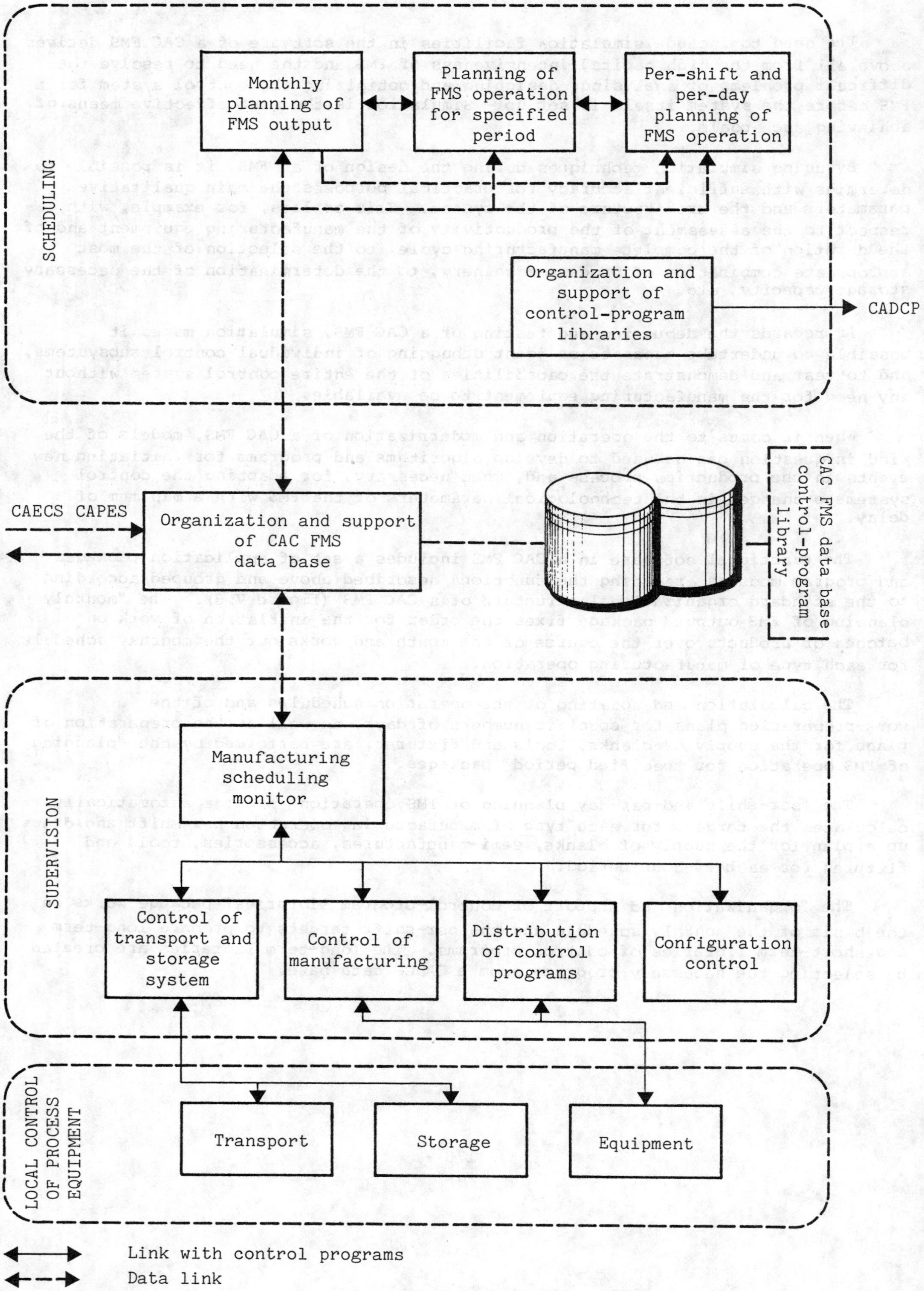

Link with control programs

Data link

The main task of the "manufacturing scheduling monitor" package is automatically to monitor and adjust the progress of manufacturing, including the supply to the manufacturing equipment of the components necessary for manufacture, and to ensure optimum loading of the manufacturing equipment in the light of the priority attached to orders and of the availability of resources for fulfilling those orders.

To perform its tasks, the package must execute the following functions:

- Compilation and management of queues of requests to individual components of the FMS;

- Reservation of the resources necessary for compliance with the priority requests;

- Harmonization and monitoring of the functioning of the "control of the transport and storage system", "control of manufacturing" and "distribution of control programs among NC-equipment" packages; and

- Recording in the CAS FMS data-base files of changes in the manufacturing process.

The package also provides a dialogue facility which enables the FMS supervisory staff to intervene rapidly in the manufacturing process when the need arises. This is helpful when the situation requires a decision concerning the future course of work (for example, a change in the sequence in which orders are to be filled, the breakdown of a machining centre or its return to service after repair or scheduled preventive maintenance, etc.).

The "control of manufacturing" package prepares the manufacturing equipment for operation, starts it up and monitors its functioning.

The preparation function includes the supply to the manufacturing area of the items (blanks, semi-manufactures, accessories, tools, fixtures, etc.) required for the next manufacturing operation and (in the absence of automatic devices for the attachment or changing of tools) the issuance of commands to the supervisory staff for the attachment or changing of tools.

The "control of the transport and storage system" package controls the movement and storage of loads within the FMS, including the movement of loads from storage points to manufacturing points and vice versa, and the monitoring of the delivery of loads to and the presence of loads in the FMS.

The function of transferring the control programs necessary for the execution of a particular processing operation by a particular piece of equipment from the working library to the local-control level is executed by the "distribution of control programs" package.

The time characteristics of software at the supervision level must meet the requirements made of real-time systems. Consequently, the time required to execute any function must not exceed the time required to control and service the manufacturing process in the FMS.

As stated above, the precise nature of the software at the local-control level depends substantially on the configuration of the manufacturing equipment in the FMS and of its local control systems. Consequently, the provision of software for this level does not lend itself to standardization.

The above-mentioned features make it necessary to develop specially for each specific FMS program modules that take into account the characteristics of the manufacturing equipment employed. The functions of such modules include the breakdown of the logic commands issued by the supervision level into a sequence of commands for the local control systems for the transport, storage and manufacturing equipment. The modules must also convert the messages received from the local control systems into messages confirming the execution of such commands.

ANNEX VI: COMPUTER-AIDED DESIGN OF MANUFACTURING SYSTEMS
SOFTWARE FOR THE MECHANICAL ENGINEERING AND
INSTRUMENT-MAKING INDUSTRIES [1]/

The problem

There is no longer any doubt of the need to automate all phases of the manufacturing cycle - research and development, product design, the development of manufacturing technology, and product manufacture and testing - within the bounds of the most important component of any production system, a computer-integrated manufacturing (CIM) system.

In its most general and fullest form, a CIM system comprises a set of harmoniously interacting local components: a computerized research system; a computerized product-design system; a computerized production engineering system; and computerized systems for the control of enterprises, works, buildings, shops and shop sections, including manufacturing processes and flexible manufacturing systems.

In a CIM system, all the above systems are interconnected and based on a common, distributed data base. The system must be capable of functioning in real time and this requirement is met by using 50 computers of various types interconnected to form a data-processing network. The computers are capable of supporting more than 300 terminals and personal computers to provide users with hard-copy printouts.

Computer-integrated manufacturing systems constitute an essentially new class of computer-aided control systems. That is so not only because their capabilities are more extensive, but also because they entail the use of what are, in essence, new data-base, software and hardware design techniques and that in turn entails the computerization and standardization of the development, debugging, testing and implementation of computer-aided control systems.

The data requirements of a CIM can be met in the form of a distributed data base. The designer of such a base has to solve the complex problems of distributing the data over control levels so as to reduce both the number of messages transmitted between levels and the duplication of calculations at different levels.

The hardware for a CIM system comprises a computer network, including mainframe computers from the ES (Unified System) series, mini- and micro-computers, work-stations, various sensors and switching devices and visual display units. For the control of manufacturing processes, the system must have a reaction time in the order of milliseconds when simultaneously processing data accumulated during a period ranging from a few seconds to

_____

[1]/ Material transmitted by the Government of the USSR.

several years and when exchanging this data between its own constituent levels. It is typical of CIM systems that, while the volume of data accumulated in the system during the control of manufacturing steadily increases, the amount of data that is input manually steadily decreases as a result of the use of automatic data recorders and the growth in the volume of control data transmitted from level to level and output on the VDUs installed at the points of use of the information.

The expansion of the range of mini- and micro-computers employed, and the differences in their basic and system software, have greatly complicated the design of software both for CIM systems and for components thereof. The designer must know several languages and programming systems as well as the peculiarities of computing on various types of machine.

The development of CIM systems requires a special approach to the securing of harmonious interaction between all the levels of such a system. This approach must take into account the distribution of data over the levels of control, the distribution of computing resources amongst the various components of the system, and the organization of the data bases and hardware configurations necessary for specific components of the system.

In this respect, there has been a growth in recent years in interest in the problems of comprehensive automation of the study, development, debugging, testing, introduction and modernization of control systems. A number of reports have been written on the creation of systems for the computer-aided design of CAM systems.

A further stage in the development of CAD-CIM systems (systems for the computer-aided design of computer-integrated systems) has been the creation of an integrated interactive system, a schematic of which appears in figure VI.1 and which operates on the basis of work-stations for designers, programmers and testers. The main objectives and principles of its design are described below.

First, the system is designed to permit the computer-aided development of CIM systems containing such functional components as systems for computerized works-management, CAD, computerized control of FMS and computerized control of manufacturing processes. Secondly, the system forms the basis for the computerization of various engineering design functions and of the development of data and software, the issuance of documentation, and the testing and debugging of programs for, and the improvement and development of CIM systems. Thirdly, the computerized functions are combined within the system on the basis of a common design data base and of technology that takes into account the peculiarities of the stages of development of all the various components of a CIM system. The elements of this CAD-CIM system operate in the interactive mode, with the designers having the leading role in the "man/machine" dialogue.

Analysis of the components of a CIM system has shown that, because of their peculiar characteristics, they must be designed by the "subsystem method" or by the "modelling method".

The subsystem method entails the development of one or more interrelated application packages capable of performing all the control functions for any given component of the CIM system. The parameters in the packages are adjusted and the reference documentation revised to suit the component.

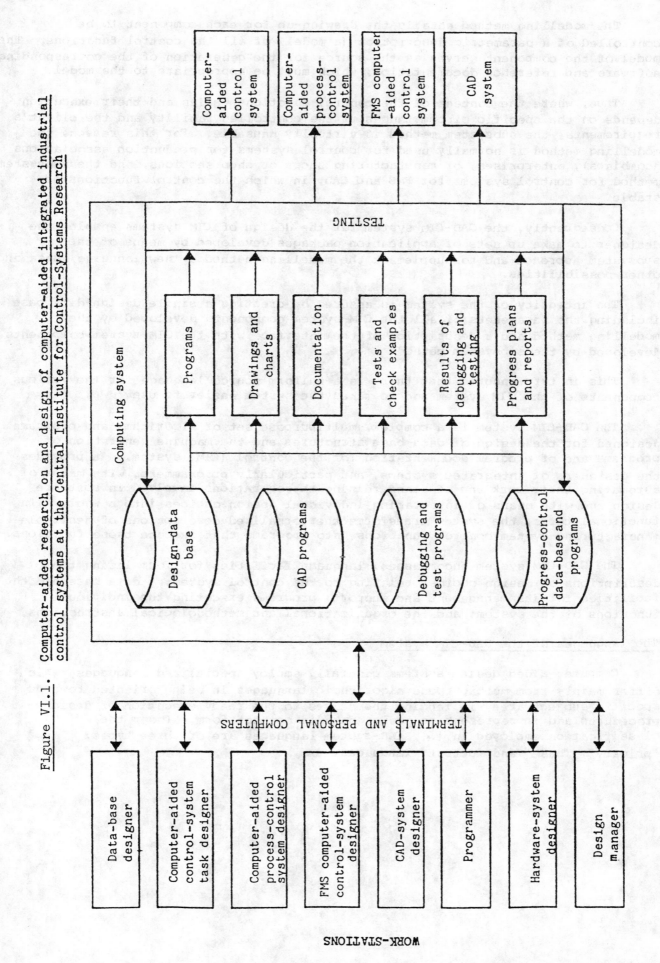

Figure VI.1. Computer-aided research on and design of computer-aided integrated industrial control systems at the Central Institute for Control-Systems Research

WORK-STATIONS

The modelling method entails the drawing-up for each component to be controlled of a parametric description (a model) of all the control functions. The model of the component serves as the source for the generation of the corresponding software and reference documentation, which must be appropriate to the model.

Thus, where the content of the control functions varies and their execution depends on the specific circumstances of the controlled facility and the client's requirements, the subsystem method is virtually unusable. For this reason, the modelling method is normally used for control systems for production associations (combines), enterprises, or manufacturing shops or shop sections, and the subsystem method for control systems for FMS and CAD, in which the control functions are stable.

Consequently, the CAD-CIM system for the design of CIM systems enables the designer to make up sets of application packages developed by means of the subsystem approach and to supplement the modelling method by new language, data and other possibilities.

The integrity of the system is ensured by creating a single design-data base including the parameters of all the CIM-system components developed by the modelling method and a description of the interface with the CIM-system components developed by the subsystem method.

This in turn ensures that there is a uniform source language for the various components of the CIM system and so makes the system easier to use.

The CAD-CIM system is a complex, multipurpose set of algorithms and programs designed for the design of data-base structures and the machine generation of programs and of program documentation for the control (CIM) system. It provides the designers of integrated systems, and particularly programmers, with means of automating their work and of standardizing the operations involved in software design and with means of integrating individual design choices into a whole. In functional terms, the system transforms the formalized descriptions of desirable manufacturing-system control functions into programs that execute those functions.

The CAD-CIM system encompasses: language facilities for formalizing the descriptions of design choices relating to the control system; a data base and the facilities for its management and support; programs executing the individual functions of the system; and the organizational and methodological instructions.

The languages of the CAD-CIM system

Computer-aided design systems generally employ specialized languages, which differ mainly from multipurpose algorithmic languages in being oriented towards a specific subject area, in lending themselves to the ready execution of design procedures and in representing design data in compact form. Under the classification employed in [5], CAD-system languages are of three types: "primary", "base" and "output" languages.

The primary languages of the CAD-CIM system are designed for formalized description of the detail design of a CIM system and permit the statement of the basic design choices. The languages of this type comprise: a language (known as ELEMENT) for describing the data processing elements and the links between them; a language (ZADACHA) for describing the the data-processing algorithms; and a language (ZAPROS) for describing inquiries for machine descriptions of data-processing elements and algorithms.

ELEMENT is a tabular language. Entries ("Rekvizity"), input and output documents, logic files, connections between documents and files and between files and tasks, etc., are described in special tables. The content and structure of the tables define the content and semantics of the descriptions; this reduces the time required for assimilation of the language by a user to the minimum and permits the rapid preparation of input descriptions. There is a particular syntax for writing each kind of table.

ZADACHA, which is a free-form language, is used for describing algorithms. The course of data processing is described with the help of this language's relatively simple operators. All the descriptions are implemented at the level of logic files. There is no definition of the physical arrangement or format of the data, but simply description of the user's requirements as regards the algorithms.

The principal advantage of ELEMENT and ZADACHA is that, by using them, the process of describing the data to be worked on can be separated from that of describing the algorithms. This makes possible: first, the preparation of the descriptions by various categories of specialist; and, secondly, the analysis of the control system for completeness and consistency of the links between its elements without the need to know the actual processing algorithms.

Apart from this, the descriptions of system elements and of the links between them are principally used in the designing of "informational" programs (programs for data input from various data carriers, data-base organization and support programs, programs for the generation of references and reports). The descriptions of the algorithms are used in the design of functional programs (programs for the joint processing of files, for the management of the computing process, etc.). Consequently, it is always possible to design CIM software through the iterative (multiple-step) process of augmenting the number of problems to be solved and improving the algorithms for their solution.

In designing systems, there is often a need for various kinds of request for the output from the data base of descriptions of individual elements or sets thereof. ZAPROS is a language specially designed for making such inquiries. It can be used to request the output from the design data base of descriptions of elements, and to inquire about the composition of elements and how they fit into or are linked with other elements. ZAPROS is a free-form language.

The base languages are principally intended for:

(a) Formalizing descriptions of the main design functions; analysis of the completeness and compatibility of the machine descriptions present in the data base and their correction as necessary; selection of the content and structure of the CIM data base; generation of informational and functional programs; machine synthesis of documentation;

(b) Optimization of design decisions: minimization of the main and external storage space required; minimization of the machine time required; choice of rational file-processing paths; timely amendment of design decisions, etc.

The base languages include: the design parameters permitting the designer to intervene as and when necessary in the decision-making process; and the intermediate languages used by the various components of the CAD-CIM system during their joint operation.

The intermediate languages are intended to ensure the harmonious interaction of functional components at the various stages of CIM-system design. They are used to formalize descriptions of intermediate results, especially: to formalize the transition from the building of a computer model of a problem to the generation of object programs; the formalization of record structure description, of the problem-solution flow-chart, etc.

The output languages are intended for outputting the design choices or, in other words, the programs and program documentation.

For informational programs the output language is ASSEMBLER; for functional programs it is COBOL.

The program documentation corresponds to the requirements of ESPD, the Unified System of Program Documentation.

The CAD-CIM system data base

The data base is the heart of the entire CAD-CIM system. It comprises the "processing model" (MO) and the "BASE MODEL" (BM).

The MO contains parametric information constituting formalized descriptions of the main elements peculiar to data collection, recording, correction, storage, processing and output in the control of a specific manufacturing system. The MO is created afresh before the start of software design for each particular CIM-system and is then used by all the functional components of the design system for the purposes of generating programs and documentation in the light of the specifics of that system. The types of element in the MO include problems, files, input documents, reports and so on. Each type of element is defined by a set of parameters. For example, each entry ("rekvizit") is defined by an identifier, a picture indicating the characteristic properties, and the check methods (by module, by range of values, by table of values, etc.).

The BM corresponds to the MO in structure, but differs from it by containing formalized descriptions of standardized forms of input and output documents and of data media for mass-produced data-preparation devices, descriptions of standard flow-charts and unified descriptions of entries ("rekvizity"). While the MO is accessible for all CAD-CIM design program packages, access to the BM is by means of the design-data-base support devices.

- 184 -

The BM is a sort of store of design solutions for use in designing systems and can serve as a source of data for the creation of an MO for a particular system.

The BM is built up by extracting the common features from design choices applied in developing CAM systems at enterprises of the same type. It is a data base, the data in which are suitable for analysis and further use by designers of new systems. The CAD-CIM system places no constraints on the contents of the BM, which can therefore be expanded and improved at any time.

The uses of an integrated design-data base in the operation of a CAD-CIM system are the following:

- Design is easier;

- There is no duplication of the information required for each component of the design system;

- The results of the design exercise are more reliable because of the comprehensive analysis and checking of the data in the base;

- Design is faster because there are no contradictions or omissions in the description of elements; and

- Control of the design process is centralized.

Software

The software in the CAD-CIM system comprises a set of application packages for the performance of the following tasks connected with the design of specific manufacturing systems in the fields of engineering and instrument-making: automation of the design of data-base structures for CAM; the design and generation of programs for the management of CIM-system data bases in the batch and conversational modes, and of programs for the input of data from various media and for the compilation and outputting of reports; the design of real-time question-and-answer systems; the design of means of managing computing in a CIM system; the compilation and output of system documentation; the generation of test (check) examples and the analysis of the quality of program testing (figure VI.2).

There follows a short description of each of the application packages.

The MARS-PROBA-KOMPLEKS, VVOD-SM-KOMPLEKS and VYVOD-SM-KOMPLEKS packages comprise a system for the generation of programs executing the following functions: organization and management of the data base for a particular integrated manufacturing system; and compilation and output of reports in a form suitable for use in the management of the manufacturing facilities concerned. Thanks to the availability of this system of packages, the algorithmization stage is completely eliminated and the functions of the informational programs are executed according to standard algorithm. Consequently, design simply requires the preparation of the following descriptions of the data to be processed: the content and structure of the input and output documents, and the methods for checking them and the data carriers; the data-base structure; the entries ("rekvizity") and the means of checking them. The packages can also be used to design file-creation

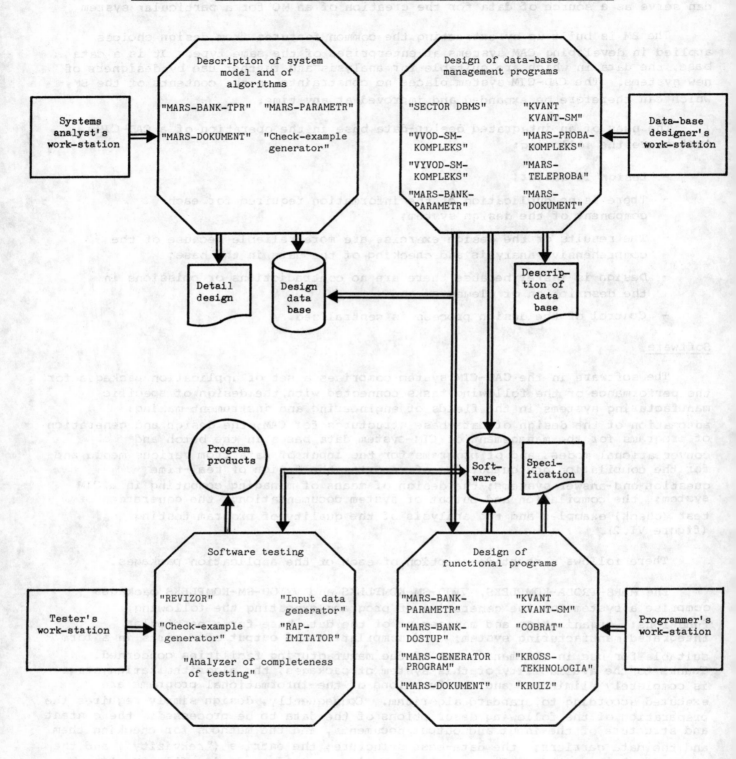

Figure VI.2. Computer-aided design of computer-integrated manufacturing systems

Description of system model and of algorithms

"MARS-BANK-TPR"   "MARS-PARAMETR"

"MARS-DOKUMENT"   "Check-example generator"

Design of data-base management programs

"SECTOR DBMS"   "KVANT KVANT-SM"

"VVOD-SM-KOMPLEKS"   "MARS-PROBA-KOMPLEKS"

"VYVOD-SM-KOMPLEKS"   "MARS-TELEPROBA"

"MARS-BANK-PARAMETR"   "MARS-DOKUMENT"

Systems analyst's work-station

Data-base designer's work-station

Detail design

Design data base

Description of data base

Program products

Software

Specification

Software testing

"REVIZOR"   "Input data generator"

"Check-example generator"   "RAP-IMITATOR"

"Analyzer of completeness of testing"

Design of functional programs

"MARS-BANK-PARAMETR"   "KVANT KVANT-SM"

"MARS-BANK-DOSTUP"   "COBRAT"

"MARS-GENERATOR PROGRAM"   "KROSS-TEKHNOLOGIA"

"MARS-DOKUMENT"   "KRUIZ"

Tester's work-station

Programmer's work-station

Key

"..."; Application package

- 186 -

programmes for the loading of a data base organized by any data-base management system (DBMS). This makes the resultant programs independent of the DBMS employed and the structure of the data base independent of the content and structure of the input documents.

The main advantage of the packages is that they provide comprehensive coverage of the design tasks involved in elaborating CIM for various types of computer. The object programs generated by the packages permit rapid changing of the input-data carriers during operation of the system and also execute various methods of checking the data entered.

The programs generated by the use of MARS-PROBA-KOMPLEKS are intended for computers in the Unified System of Computers running under that system's own specially-designed OS ES operating system. The programs developed with the help of VVOD-SM-KOMPLEKS and VYVOD-SM-KOMPLEKS are intended for the SM-1420 machine and the OS RV operating system.

The automated design of inter-active control systems for teleprocessing of data in CIM is accomplished by means of the MARS-TELEPROBA package. This package includes a suite of interrelated algorithms and programs for the real-time collection of data from remote and local terminals and for the interactive updating of the data base.

Real-time operation is a means of making efficient use of computers: it enables a number of users, each with his own terminal, to input data at its point of origin and to obtain up-to-the-minute information on the status of production. Real-time systems do, however, have a number of features that seriously complicate their design. The development of software for such systems is particularly difficult. MARS-TELEPROBA includes the following program components: a versatile system for the control of real-time computing in multi-level CAM Systems with remote data-acquisition and -display terminals (the KVANT application package); a system for machine design of application programs for interactive updating of the data base, real-time processing of inputs to the computer system and real-time functioning of question-and-answer systems.

The SETOR DBMS is designed for the building and maintenance of CIM data bases with a net-type structure. The package provides the user with languages for the description and manipulation of data and with means of reorganizing and updating the data base. The package supports the logic links between files and between file records. SETOR ensures the independence of data at the field level, data integrity and reliability, and the protection of data against unauthorized access. The package is simple to understand and use, provides rapid access to data, requires little space in main memory and offers a wide range of service facilities.

One of the most important areas of work as regards the automation of software design for computer-aided control systems is that of the creation of means for the design of data-base structures.

The STRUKTURA-KOMPLEKS package contains algorithms and programs for designing the composition and structure of a CIM data base. It permits the design of: rational, two-level net-type data structures compatible with the requirements of the SETOR DBMS; and separate files and means of arranging them in external store. The package operates in the batch and interactive modes and the user can intervene in the decision-selection process in each mode with the help of special design parameters. As a result, the package can be used both in the initial design of a data base and in the amendment of an existing structure for the purposes of modernizing a computer-aided control system.

The source data for the package is in the form of formalized descriptions of the list and characteristics of the tasks that interact with the base being designed. Here are defined the form of entries ("rekvizity"), the structure, and the space-time characteristics and desired types of access to the logic files for use in the tasks.

The MARS-GENERATOR PROGRAMM package is intended for the machine design and generation of programs and of means of starting and executing those programs in accordance with the processing flow chart. In other words, the package transforms algorithms for the solution of CIM problems coded in the ZADACHA language into programs that will execute those algorithms. MARS-GENERATOR PROGRAMM provides algorithms for data-processing operations. In particular, there is automatic determination of the need for the reordering of input and intermediate files, of the possibility of using a computer's working memory for processing, and of the most efficient method of merging, etc. The results of the use of the package are modules in COBOL for the performance of specific tasks and a control program. The control program receives the operating parameters, transfers them to the processing modules, stores the values of the parameters in the processing modules and stores the values of the parameters in the work file for the task or the CIM subsystem.

The software generated by means of the package is independent of the physical structure of the CIM data base. This is possible because the modules that are generated call on the MARS-BANK-DOSTUP package for all reading, location or writing of data in the data base.

MARS-GENERATOR PROGRAMM operates on computers of the Unified System in a OS ES environment. Programs generated by the package can be run both on machines of the Unified System and, in an OS RV environment, on computers in the SM series.

The MARS-BANK-DOSTUP package provides, on the basis of parametric descriptions, access to the CIM data base; that access is independent of the structure of the base. The CIM data base can contain files for the OS ES and OS RV operating systems and/or the SETOR DBMS. Problem programs running in any language that contains the operator CALL can address MARS-BANK-DOSTUP with a request for logic files and the package will instantaneously determine what physical files in the data base should be called to meet the request.

All the packages that have been described above are integrated on the basis of common data stored in the CAD-CIM system data base. Machine creation of high-quality operating documentation for the system under design can therefore be undertaken whenever required. At any given time in the design process, the CAD-CIM data base will contain up-to-date, complete descriptions of all the data-processing elements of the CIM system in question. The principal function of the MARS-DOKUMENT package is to convert the coded descriptions of CIM-software design solutions generated by the CAD-CIM system into the requisite operating documents. The main advantage of using MARS-DOKUMENT in CIM-system design is that, in order to produce the documentation, the designers do not have to supply any data beyond what is already present in the design-data base.

From the functional viewpoint, the CAD-CIM data base is a specialized base with its own management system and contains all the information necessary for the machine generation of CIM software. Management of the base is affected by the MARS-BANK-TPR package, which contains programs for organizing and maintaining the base and the means of analysing and accessing the data in it.

MARS-BANK-TPR has a number of operating features that distinguish it from the other packages in the CAD-CIM design system. In particular, it contains no provision for adjustment to the parameters of the CIM system being designed. MARS-BANK-TPR is a tool for the automatic integration within the data base of the descriptions of the elements of the particular design in question.

The programs in the CAD-CIM system are modular. This means that any of the application packages can be deleted or replaced without diminishing the conceptual integrity or the capability of the system. For example, should the STRUKTURA-KOMPLEKS package not be used during the logic design of the data base of a particular system, the elected data-base structure can be described using the CAD-CIM system languages. The CAD-CIM system programs can load these descriptions into the design-data base for subsequent use in the development of programs and documentation for the CIM system in question.

1. **System MOSY (developed by Microsystem-Sztaki/Hungary)**

MOtion SYnthetizer (MOSY) is a software tool which enables the user to create and display computer generated stationary and moving pictures based on two- and three-dimensional geometric models.

MOSY can be used in:

- The development of computer animated three-dimensional motion sequences (movies);

- The creation and graphic presentation simulations of robot programs;

- The simulation of robot arm motion and the manufacturing environment;

- Architectural and ergonomic design; and

- Any other task where one wants to create computer generated series of images based on geometric models.

The MOSY three-dimensional animation system consists of two basically different program parts. There is an interpreter which performs the motion simulation itself and there is another - interactive - program which supports completion of the necessary preparation tasks (e.g. assigning reference frames to moving objects, setting camera parameters, etc.).

Motion simulation

The MOSY programming language developed for describing the "script" of motion sequences of several "actors", incorporates the usual computer language statements (variable declarations, control structures, expressions, etc.) as well as a number of others related to spatial animation. The special statements help the user in constructing a "scene" and in moving objects on that scene.

As the task of the system is to simulate spatial motion, special data types have been introduced to support the handling of three-dimensional data.

ASSEMBLY 2/

An ASSEMBLY (a collection of objects to be moved together) is essentially a REFERENCE co-ordinate system and a set of MODELs and other ASSEMBLY type objects associated with it. It can be imagined as a tree structure, where every node is an ASSEMBLY and every leaf a MODEL.

1/ Extracted from information provided by MICROSYSTEM-SZTAKI, Hungary.

2/ The meaning of certain words written in capitals is contained in the glossary at the end of this section.

SCENE

Motion simulation takes place in a co-ordinate system called SCENE. Motion simulation is controlled by a user-written program called SCENE-PROGRAM.

At the very beginning the user has to establish the conditions governing his expected animation sequence. It means that a SCENE has to be built up from existing MODELS and ASSEMBLIES, which will be ACTORS (if they will move) or parts of the "background" (if they will be stationary).

CAMERA

All the motions are displayed by a virtual CAMERA. Since the CAMERA is treated as an ordinary ACTOR a motion program called CAMERA-PROGRAM can be assigned to it. This means that one can create images in which only the CAMERA moves around or its other features change.

Objects can be displayed with hidden lines removed and shading can be incorporated into the system by using a suitable geometric modeller.

ROBOT

The ROBOT is an ASSEMBLY and a ROBOT actor can be moved by a ROBOT-PROGRAM. Special robot motion commands are part of the MOSY three-dimensional animation language, and Cartesian-to-joint transformation modules for user-specified robot arms can easily be inserted into the system.

Interactive support

The interactive part of the animation system helps the user

- In creating the reference frame of a SCENE or that of an ASSEMBLY-PROGRAM;

- To complete a MODEL with some additional data, e.g. model references; and

- To control the process of animation (e.g. stepwise execution, run with selected actors only, making selected ACTORS invisible for a while, changing workspace references, replay by wireframe, hiddenline, "stick" or "brick" representation, etc.).

From each menu of the interactive part of MOSY, a CAMERA menu can be activated where the position, orientation and other features of the virtual camera can be set and tuned. Using the CAMERA menu, the user can examine his workspace (the scene) from any viewpoint.

COMPLETER

Since motion simulation may call for some information which may not be available in the data base of the chosen geometric modeller, there is a menu called COMPLETER where these missing data can be specified. In the COMPLETER the user can create REFERENCEs, STICK models, and BRICK approximation of objects.

DIRECTOR

This menu is the user interface to the motion simulation. Before the system performs a SCENE-PROGRAM, the user can select by option switches the required kind of animation.

Though the DIRECTOR has many options, only one of them is mentioned here, namely collision detection. Collision detection means that the system controls the position of a selected set of ACTORs in every step (phase) of the animation and if a possible collision is detected, the user is warned.

Glossary

ACTOR   can be a MODEL or an ASSEMBLY. If a MODEL or an ASSEMBLY has to be moved in the SCENE, it is necessary to assign to it a so-called ACTOR-PROGRAM.

ACTOR-PROGRAM   is responsible for one kind of motion of an ACTOR at a given instant, so if a combined motion (the net motion of several elementary translations and rotations as well as cyclic motion components) is needed, several ACTOR-PROGRAMS have to be assigned to one ACTOR.

ASSEMBLY   is a co-ordinate system which may contain MODEL(s) and ASSEMBLY(ies). An ASSEMBLY can be moved as one entity. This can also be done in such a way that only some of its component elements change their position and each of them can move along a different path.

ASSEMBLY-
PROGRAM   describes how the given ASSEMBLY is constructed from MODEL(s) and ASSEMBLY(ies) and what their permitted motions are. The commands to be used in an ASSEMBLY-PROGRAM are almost the same as those of the SCENE-PROGRAM.

MODEL   is an elementary geometric model generated by a geometric modeller (e.g. MODBUILD).

POINT   is a special data type in the system. It can mark a three dimensional point in any of the three kinds of co-ordinate systems (SCENE, ASSEMBLY, MODEL).

REFERENCE   is a special data type in the system. It can mark a co-ordinate system in any of the three kinds of co-ordinate systems (SCENE, ASSEMBLY, MODEL).

SCENE   is a co-ordinate system in which all the participating geometric objects (ACTORs, POINTs, REFERENCEs, etc.) find their place during motion simulation.

SCENE-PROGRAM describes the entire animation. It can contain statements for
designing the layout of the whole SCENE as required, for assigning
ACTOR programs to MODEL and ASSEMBLY type objects as well as all the
required ACTOR-PROGRAMS.

2. Programming system MODBUILD

Fundamentals

MODBUILD is a programming system with a sophisticated human interface. Its
purpose is to build, change (edit) and store three-dimensional objects as well as
to compose new objects from those already composed. The objects are in fact
geometric data structures, i.e. geometric models. These models can be used for
several purposes by external programs (e.g. 3D animation, fixture design, robot
modelling, etc.). MODBUILD displays them as three-dimensional objects in a
two-dimensional plane (i.e. on the screen). This is done via perspective or
axonometric projection. The parameters of the projection can be defined (and/or
changed) by the user himself. A special feature of MODBUILD is its "hidden line"
algorithm. This makes it possible to display the model in such a way that the
lines, hidden by a surface, are not displayed, thus allowing a much more
spectacular view of the objects.

Human interface of MODBUILD

A sophisticated human interface has two duties to fulfil:

- It has to provide a powerful set of input tools in a user-friendly environment
  (i.e. easy usage, appropriate prompts, etc.); and

- It has to provide an immediate and "easy to understand" feedback on what is
  going on, i.e. the consequence of the operator actions.

MODBUILD provides effective solutions for both these problems. The system
should contain a graphic terminal with a keyboard and a "locator" type peripherial
device (graphic cursor).

. MODBUILD works in four modes:

- Load-save mode

- Viewing mode

- Modelling mode

- Service mode

In the load-save mode, one can load in the already created models for further
manipulation and save out (retain) just-created new models or parts of them. In
the viewing mode the user can visually check the models from an arbitrary viewpoint
and/or in an arbitrary projection direction. In the modelling mode models can be
built, i.e. model elements can be generated, deleted or changed. In the service
mode the system parameters can be maintained.

- 194 -

The geometric objects in MODBUILD

A geometric object can be considered mathematically as a tree. The top element of the tree is the body. A body is composed of surfaces. A surface is bordered by edges. An edge can be:

- A straight line; or

- A circular arc.

The internal structure of MODBUILD

MODBUILD consists of four subsystems which can be developed and modified almost independently from each other. This gives MODBUILD a high degree of flexibility, i.e. it can be tuned for speed and space, and special user requests can be easily satisfied.

The subsystems are the following:

- The menu-handling subsystem controls the non-machine interaction;

- The data-structure subsystem supervises the data management;

- The display subsystem is responsible for the displaying of the three-dimensional objects; and

- The hidden line subsystem provides the hidden line algorithm.

MODBUILD is implemented in portable C language, i.e. it is easy and fast to reimplement on different computers.

ANNEX VIII: PROGRAM SYSTEMS FOR CLASSIFICATION AND GROUP TECHNOLOGY 1/

The processes of design, production planning and transfer to series production give rise to high expenses which for a single component may attain some 1,500 leva. For wide assortments of workpieces, it is therefore appropriate to use manufacturing documents already elaborated for similar parts.

In this context, a computer-aided system has been developed in Bulgaria for coding and classifying workpieces by the group technology system.

The system consists of the following components:

- DIACLASS;

- DIAGRUP;

- DIATECH.

Before describing these components in detail, an indication of their functioning seems in order.

By means of DIACLASS classification and coding, it is possible to determine which parts are similar from the point of view of design and technology. Thus the repeated design of similar workpieces can be avoided. DIACLASS is a dialogue-oriented system searching for similar parts based on qualitative and quantitative criteria which may be proposed by the user.

Using the coding results of DIACLASS, the component DIAGRUP carries out a grouping of parts by parameters either system-internal or user-oriented (i.e. user-proposed). While menus are offered directed to special strategies, the user is also able to make his own menus to accommodate specific subjects. The Menus can be stored and accessed. Thus the principle of group technology can be applied to engineering.

DIATECH is a system for the automated computer-aided design of technological documentation using as the codes resulting from DIACLASS as functions of DIAGRUP. It is possible to elaborate work sheets for manufacturing with the help of DIATECH.

1.   DIACLASS - an automated dialogue-oriented system for coding and classification

DIACLASS is based on the method of searching for and compiling similarities.

The code in the DIACLASS system represents the key for the structuring and accessing of the data base needed for computer-aided production planning, analysis for group technology and other activities connected with CAD/CAM.

_____

1/ Extracted from Bulgarian sources.

The system is usually delivered with already prepared codes for parts to be machined. Decision tables and other methods, however, enable the user to undertake his own coding where necessary.

The subsystem for parts to be manufactured records the following information about the workpieces:

- Main shape;

- Dimensions, sizes;

- Tolerances;

- Shape and dimensions of the unmanufactured part; and

- User-specific data.

The programs for classification and coding can be used in an interactive mode. By using the DIACLASS principles of group technology it is possible to subdivide the parts into groups of high similarity based on aspects of design, technology, manufacturing and manufacturing organization.

DIACLASS is implemented on computers such as PDP 11, SM - 4 and BK - 1302. It requires a main storage capacity of 128 KByte.

2. DIAGRUP - an automated dialogue-oriented system for grouping

DIAGRUP is used for the analysis of production assortments on the basis of various criteria. These criteria are established by the data stored in the DIACLASS system.

The system undertakes a search in accordance with the user's criteria. The output data are structured accordingly.

The grouping itself is carried out by the user following a chosen strategy. Some objectives of grouping could be as follows:

- Aspect of unique designed shapes (part geometry);

- Grouping following quantity parameters;

- Analysis of various types of operations;

- Balancing of demands against the capacity of manufacturing equipment or department; and

- Grouping of parts having the same sequence of operations with the aim of optimizing the manufacturing organization.

Optional strategies advantageous from the point of view of the user can be stored.

The system has been installed for micro- and minicomputers having a main storage capacity of 64 to 128 KByte.

The structure of DIAGRUP is based on the same principles as DIACLASS. Both components can be linked. Thus they can be esteemed to represent a unique program system for the demands of production planning, design and manufacturing.

REFERENCES

[1] Production and Use of Industrial Robots. United Nations Economic Commission for Europe, New York 1985. (United Nations publication, Sales No. E.84.II.E.33).

[2] Recent Trends in Flexible Manufacturing. United Nations Economic Commission for Europe, New York 1986. (United Nations publication, Sales No. E.85.II.E.35).

[3] Software: An Emerging Industry. Organisation for Economic Co-operation and Development, Paris 1985.

[4] Proceedings of the 1st International Conference on Flexible Manufacturing Systems, 20-22 October 1982, Brighton, United Kingdom (IFS Publictions Ltd.).

[5] Proceedings of the 2nd International Conference on Flexible Manufacturing Systems, 26-28 October 1983, London, United Kingdom (Bedford, United Kingdom, IFS Publications; Amsterdam, North-Holland Publishing Company).

[6] Proceedings of the 3rd International Conference on Flexible Manufacturing Systems and 17th Annual IPA Conference, 11-13 September 1984, Boeblingen, Federal Republic of Germany, (Bedford, United Kingdom, IFS Publications; Amsterdam, North-Holland Publishing Company).

[7] Conference Proceedings of AUTOFACT III, 9-12 November 1981, Detroit, United States, (United States, Society of Manufacturing Engineers).

[8] Conference Proceedings of AUTOFACT EUROPE, 13-15 September 1983, Geneva, Switzerland, (United States, Society of Manufacturing Engineers).

[9] "Factory of the Future", Industry Week, 1984 (Date unknown).

[10] "Factory of the Future", Industry Week, 4 March 1985.

[11] International Standard ISO 7498, International Organization for Standardization, 1984.

[12] International Standard ISO 2382/1, International Organization for Standardization, 1984.

[13] A. Chandor, The Penguin Dictionry of Microprocessors, Penguin Books Ltd., 1981.

[14] Chr. Rockrohr, Wörterbuch der Daten- und Telekommunikation, Markt und Technik, München, 1983 (In German only).

[15] International Standard ISO 2382/9, International Organization for Standardization, 1984.

[16] A Competitive Assessment of the US Manufacturing Automation Equipment Industries, US Department of Commerce, 1984.

[17] Draft International Standard ISO/DIS 2382/18, International Organization for Standardization, 1984.

[18] Numerical control of machines, ISO Standards Handbook 7, International Organization for Standardization, 1981.

[19] M. Hordeski, Illustrated Dictionary of Microcomputer Terminology, TAB Books Inc., 1978.

[20] Proceedings of the UN/ECE Seminar on Flexible Manufacturing Systems: Design and Applications, 24-28 September 1984, Sofia, Bulgaria (UN/ECE Working Party on Engineering Industries and Automation).

[21] "CASA Wheel Revised", CIM Technology, Spring 1986.

[22] "Local Area Networks - The Future of the Factory", Manufacturing Engineering, March 1986.

[23] D. Kochan, "Development in Computer-Integrated Manufacturing", IFIP - State of the Art Report - CAM, Springer-Verlag Berlin - Heidelberg - New York - Tokyo, 1986.

[24] Document LPCS/WGTQ/I/1-3, Working Group on Technical Questions Relating to the Legal Protection of Computer Software, World Intellectual Property Organization (WIPO), Geneva, 1984.

[25] Electronic Business, 15 September 1986.

[26] M. Vossloh, "Breites Spektrum in der Entwicklung numerischer Steuerungen" (A wide spectrum of the development of numerical controls) Werkstatt und Betrieb, 116 (1983) 9 (In German only).

[27] H. Wörn, Numerische Steuersysteme - Aufbau und Schnittstellen eines Mehrprozessorsteuersystems (Numerical control systems - structure and interfaces of a multi-processor-control system), ISW 27, Springer-Verlag Berlin - Heidelberg - New York, 1979 (In German only).

[28] U. Spieth, Numerische Steuersysteme - Hardwareaufbau und Ablaufsteuerung eines Mehrprozessorsteuersystems (Numerical control systems - hardware structure and sequencing of a multi-processor control system), ISW 40, Springer-Verlag, 1982 (In German only).

[29] D. Plasch, Numerische Steuersysteme - Standardisierte Softwareschnittstellen in Mehrprozessorsteuersystemen (Numerical control systems - standardized software interfaces in multi-processor control systems), ISW 46, Springer-Verlag, 1983 (In German only).

[30] P. Klemm, <u>Strukturierung von flexiblen Bediensystemen für numerische</u> <u>Steuerungen</u> (Structuring of flexible operating systems for numerical controls), ISW 49, Springer-Verlag, 1984 (In German only).

[31] K. Holtz, <u>Multitasking-Betriebssystem für eine Mehrprozessorsteuerung MPST</u> (A multitasking operating system for a multi-processor control MPST), HGF-Bericht 84/16 (In German only).

[32] K. Holzt, B. Walker, <u>Generator für Funktionssoftware von numerischen</u> <u>Steuerungen</u> (A generator for functional software of numerical controls), HGF-Bericht 84/28 (In German only).

[33] E. Kehrer, "Zur Strukturierung von Software-Systemen in CNC-Steuerungen" (On the structuring of software systems in CNC controls), Doctoral thesis, Karl-Marx-Stadt, Institute of Technology, 1985 (In German only).

[34] T. Moriwaki, "Sensing and prediction of cutting tool failure", <u>Bulletin of the</u> <u>Japan Society of Precision Engineering</u>, Volume 18, number 2, June 1984.

[35] <u>Verfahren und Vorrichtung zur automatischen Freistellung und Rückführung eines</u> <u>Werkzeugs einer Werkzeugmaschine in bezug auf ein zu bearbeitendes Werkstück</u> (Technique and device for the automatic releasing and returning of a tool of a machine tool with respect to a workpiece to be machined), B 230 15/00, 3126 276 (Federal Republic of Germany) (In German only).

[36] <u>Method of numerical control and device therefor</u>, EP G05 B 19/403 46 032.

[37] <u>Verfahren zum Wiederanfahren eines Werkzeugs an eine Werkstückkontur</u> (A technique for the re-positioning of a tool on to a workpiece contour), G05 B 19/18 3311 119 (Federal Republic of Germany) (In German only).

[38] B. Schennerlein, "Grundlagen zur funktionellen Gestaltung numerischer Steuersysteme für die bedienerlose Fertigung und Entwicklung eines multivalent nutzbaren Softwarebausteins" (Fundamentals for the functional design of numerical control systems for the unattended manufacturing and the development of a multivalently usable software module), Doctoral thesis, Dresden University of Technology, 1986 (In German only).

[39] G. Ehrecke, <u>Planning of manufacture and equipment in enterprises</u>, ZWF 78 (1984) 12, Carl-Hanser-Verlag, München, 1984.

[40] W. v. Zeppelin, W. Klauss, <u>Progress in the development of CNC controls for</u> <u>turning machines</u>, ZWF 79 (1984) 6, Carl-Hanser-Verlag, München, 1984.

[41] H. Meier, <u>Development of a universal CNC system to obtain machine related</u> <u>automation functions</u>, ZWF 79(1984) 08, Carl-Hanser-Verlag, München, 1984.

[42] A. Potthast, P. Kobe, <u>Situation and development of CNC controls</u>, Report of the IMTS '84 in Chicago, ZWF 79 (1984) 12, Carl-Hanser-Verlag, München, 1984.

[43] A. Potthast, H. Hammer, <u>Graphical-dynamical simulation for drilling and cutting</u>, ZWF 80 (1985) 09, Carl-Hanser-Verlag, München, 1985.

[44] I. Bauer, "Sensors for an AC-limit regulation in deep drilling", <u>VDI-Zeitschrift</u> 128 (1986) 4.

[45] Vennewarld, Olde, Kesselberg, <u>Jig boring in production with FMS</u>, ZWF 79 (1984) 12, Carl-Hanser-Verlag, München, 1984.

[46] Y. Ito, <u>Present and future trends of FMS in Japan</u>, ZWF 80 (1985) 3, Carl-Hanser-Verlag, München, 1985.

[47] A. Potthast, E. Hohwieler, <u>Graphical-dynamical simulation for the CNC - double carriage working</u>, ZWF 80 (1985) 8, Carl-Hanser-Verlag, München, 1985.

[48] K. Rudolph, W. Stanek, <u>Requirements for simulation systems</u>, ZWF 79 (1984) 4, Carl-Hanser-Verlag, München, 1984.

[49] M.E. Duncan, D.I. Williams, "Software tools for automated manufacturing cells", <u>UN/ECE Seminar on Industrial Robotics '86 - International Experience, Developments and Applications</u>, Brno, 24-28 February 1986.

[50] <u>A Competitive Assessment of the US Flexible Manufacturing Systems Industry</u>, US Department of Commerce, 1985.

[51] G. Spur, Furgas, et al., "How and where industry uses its robots", <u>The Industrial Robot</u>, March 1985.

[52] "Designed to deliver", <u>Engineering</u>, May 1985.

[53] J. Osterhus, "Material Handling: A Technology of Productivity", <u>Business Week</u>, 21 January 1985.

[54] K. Yamazaki, H. Suzuki, "Systematic technology of Mechatronics Engineering and Education", in <u>Robotics and Factories of the Future</u>, Springer-Verlag, Berlin - Heidelberg - New York - Tokyo, 1984.

[55] IBM-Nachrichten, (1985) 10.

[56] G. Haarmann, "Softwaretrends", in <u>Bit</u> (1986) 1, S. 42 - 45.

[57] Model provision on the protection of Computer Software, Genf, 1978, WIPO.

[58] D. Kochan et al., <u>CAD/CAM-Schlüsseltechnologie und Intensivierungsfaktor</u>, VEB Verlag Technik, Berlin, 1986.

[59] P. Eustace, "The possible dream", <u>The Engineer</u>, 23 May 1985.

[60] Ch. Powley, "A depth of experience unique in UK", <u>Advanced Manufacturing Technology</u>, July 1985.

[61] W. Eversheim, J. Schulz, "Survey of Computer Aided Process Planning Systems", *Annuals of the CIRP*, Vol. 34/2/1985.

[62] *Advanced Manufacturing Equipment in the Community*, Commission of the European Communities, Brussels, 1985 (COM(85) 112 final).

[63] Communication of the Government of Czechoslovakia to the thirteenth session of the Senior Advisers to ECE Governments on Science and Technology, September 1985 (SC.TECH/R.178/Add.3).

[64] "Japan's $100 m software boost", *Financial Times*, 18 September 1985.

[65] "Alvey's joint approach", *Engineering*, January 1985.

[66] *Financial Times*, 9 August 1984.

[67] *Proposal for a council decision adopting the first European Strategic Programme for Research and Development in Information Technologies* (ESPRIT), Brussels, 2 June 1983 (COM(83) 258 final).

[68] "EEC project focuses on CIM", *American Machinist*, April 1985.

[69] "Partnership to improve software", *Financial Times*, 15 November 1985.

[70] "Pushing the state of the art", *Datamation*, 1 October 1985.

[71] "Erste Erfahrungen mit ESPRIT", *Neue Zürcher Zeitung*, 18 December 1985.

[72] "Projects endorsed for EUREKA", *Financial Times*, 7 November 1985.

[73] "EUREKA: one step forward", *TECH-EUROPE*, 2/1986.

[74] *Comprehensive Programme to Promote the Scientific and Technological Progress of the Member Countries of the Council for Mutual Economic Assistance up to the Year 2000*, CMEA secretariat, Moscow, 1986.

[75] K. Owen, *Computers in industry*, IFIP secretariat, December 1984.

[76] *IFAC Information - Aims, Structure, Activities*, IFAC secretariat, 1984.

[77] *ISO Memento 1986*, ISO secretariat, 1986.

[78] *Annual Report 1984*, International Electrotechnical Commission, 1985.

[79] *The IEC and Information Technology*, International Electrotechnical Commission, 1985.

[80] "Eureka go-ahead for 62 projects worth $2.1 bn", *Financial Times*, 1 July 1986.

[81] Document UNESCO/WIPO/GE/CCS/3, Group of Experts on the copyright aspects of the protection of computer software, Geneva, March 1985.

[82] <u>Invention Management</u>, March 1985.

[83] <u>A Competitive Assessment of the US Software Industry</u>, US Department of Commerce, 1984.

[84] G. Hutchinson, "Information – the unifying force in production systems", APMS – Compcontrol Conference, Budapest, 1985.

[85] D. Gayman, "Ramping up to automation with machining centres", <u>Manufacturing Engineering</u>, October 1986.

[86] <u>Business Week</u>, 21 January 1985.

[87] Proceedings of the IFIP Working Conference "Off-line programming for industrial robots", Stuttgart, June 1986.

[88] Proceedings of the IFIP PROLAMAT conference, Paris, June 1985.

[89] "Putting MAP to work", <u>American Machinist and Automated Manufacturing</u>, January 1986.

[90] E. Yourdon, L. L. Constantine, "Structured design, fundamentals of a discipline of computer program and systems design", Prentice Hall, Englewood Ciff, N.J., 1979.

[91] D.T. Ross, "Structured analysis (SA)", IEEE, Transactions of software engineering, Vol. SE-3, No. 1, 1977.

[92] GMD – Bericht, No. 181, 1980.

[93] "CAD/CAM systems", Japan Machinery Importers Association, Tokyo, 1985.

[94] Flexible Manufacturing Systems from WMW. Machine Tool and Tool Industry of the German Democratic Republic (pamphlet).

[95] "FINPRIT – The Finnish Programme for Research and Development in Information Technologies", <u>SAHKO – Electricity and Electronics,</u> 6 June 1986.

[96] <u>Annual Review of Engineering Industries and Automation, 1985</u>, United Nations Economic Commission for Europe, New York 1987. (United Nations publication, Sales No. E.86.II.E.30).

[97] IBM Manufacturing Systems as part of Computer Integrated Manufacturing, IBM Deutschland GmbH (pamphlet).

ECE PUBLICATIONS ON ENGINEERING INDUSTRIES AND AUTOMATION

Sales publications */

Digital Imaging in Health Care (ECE/ENG.AUT/25), Sales No. E.86.II.E.29,
New York, 1987. $48.00.

Recent Trends in Flexible Manufacturing (ECE/ENG.AUT/22), Sales No. E.85.II.E.35,
New York, 1986. $33.00.

Production and Use of Industrial Robots (ECE/ENG.AUT/15), Sales No. E.84.II.E.33,
New York 1985. $25.00.

Measures for Improving Engineering Equipment with a view to More Effective Energy
Use (ECE/ENG.AUT/16), Sales No. E.84.II.E.25, New York 1984. $16.50.

Engineering Equipment and Automation Means for Waste-Water Management in ECE
Countries, vols. I and II (ECE/ENG.AUT/18), Sales Nos. E.84.II.E.13 and
E.84.II.E.23, New York 1984. $12.50 (vol.I) and $8.50 (vol.II).

Techno-Economic Aspects of the International Division of Labour in the Automotive
Industry (ECE/ENG.AUT/11), Sales No. E.83.II.E.14, New York 1983. $23.00.

Development of Airborne Equipment to Intensify World Food Production
(ECE/ENG.AUT/4), Sales No. E.81.E.24, New York 1981. $16.00.

Annual Reviews of Engineering Industries and Automation

1985 (ECE/ENG.AUT/26), Sales No. E.86.II.E.30. $38.00 (vols.I and II)
1983-1984 (ECE/ENG.AUT/19), Sales No. E.85.II.E.43. $27.00 (vols.I and II)
1982 (ECE/ENG.AUT/17), Sales No. E.84.II.E.12. $11.00
1981 (ECE/ENG.AUT/10), Sales No. E.83.II.E.20. $11.00
1980 (ECE/ENG.AUT/7), Sales No. E.82.II.E.18. $13.50
1979 (ECE/ENG.AUT/3), Sales No. E.81.II.E.16. $8.00

Bulletins of Statistics on World Trade in Engineering Products

1985, Sales No. E/F/R.87.II.E.10. $45.00
1984, Sales No. E/F/R.86.II.E.10. $38.00
1983, Sales No. E/F/R.85.II.E.11. $35.00
1982, Sales No. E/F/R.84.II.E.5. $38.00
1981, Sales No. E/F/R.83.II.E.8. $38.00
1980, Sales No. E/F/R.82.II.E.5. $26.00
1979, Sales No. E/F/R.81.II.E.13. $26.00

---

*/ Sales publications and documents out of print may be requested in the form of
microfiches.

Price per microfiche $1.65; printed on paper $0.15 per page.

Documents issued in mimeograph form

Reports of Seminars held under the auspices of the ECE Working Party on Engineering Industries and Automation as from 1981:

Seminar on Industrial Robotics '86 - International Experience, Developments and Applications, Brno, Czechoslovakia, 24-28 February 1986 (ENG.AUT/SEM.5/4).

Seminar on the Development and Use of Powder Metallurgy in Engineering Industries, Minsk, Byelorussian SSR, 25-29 March 1985 (ENG.AUT/SEM.4/3).

Seminar on Flexible Manufacturing Systems: Design and Applications, Sofia, Bulgaria, 24-28 September 1984 (ENG.AUT/SEM.3/4).

Seminar on Innovation in Biomedical Equipment, Budapest, Hungary, 2-6 May 1983 (ENG.AUT/SEM.2/3).

Seminar on Present Use and Prospects for Precision Measuring Instruments in Engineering Industries, Dresden, German Democratic Republic, 20-24 September 1982 (ENG.AUT/SEM.1/3).

Seminar on Automation of Assembly in Engineering Industries, Geneva, Switzerland, 22-25 September 1981 (AUTOMAT/SEM.8/3).